MONTGO
ROCKVIL
ROCKVIL P9-CEK-949

ORGANIC CHEMISTRY AS A SECOND LANGUAGE

DR. DAVID R. KLEIN

Johns Hopkins University

JOHN WILEY & SONS, INC.

308897

AUG 1 0 2005

Acquisitions Editor	Deborah Brennan
Marketing Manager	Robert Smith
Production Manager	Pamela Kennedy
Production Editor	Sarah Wolfman-Robichaud
Cover Designer	Madelyn Lesure
Illustrations Editor	Anna Melhorn

This book was set in 10/12 Times Roman by Matrix Publishing Services and printed and bound by Courier/Westford. The cover was printed by Phoenix Color.

Recognizing the importance of preserving what has been written, it is a policy of John Wiley & Sons, Inc., to have books of enduring value published in the United States printed on acid-free paper, and we exert our best efforts to that end.

Copyright © 2004 John Wiley & Sons, Inc. All rights reserved.

No part of this publication may be reproduced, stored in a retrieval system, or transmitted in any form or by any means, electronic, mechanical, photocopying, recording, scanning or otherwise, except as permitted under Sections 107 or 108 of the 1976 United States Copyright Act, without either the prior written permission of the Publisher, or authorization through payment of the appropriate per-copy fee to the Copyright Clearance Center, Inc. 222 Rosewood Drive, Danvers, MA 01923, (978)750-8400, fax (978)750-4470. Requests to the Publisher for permission should be addressed to the Permissions Department, John Wiley & Sons, Inc., 111 River Street, Hoboken, NJ 07030, (201)748-6011, fax (201)748-6008, E-mail: PERMREQ@WILEY.COM.
To order books or for customer service please call 1-800-CALL WILEY (225-5945).

ISBN 0-471-27235-3

Printed in the United States of America

10 9 8 7 6 5

INTRODUCTION

HOW TO USE THIS BOOK

Is organic chemistry really as tough as everyone says it is? The answer is yes and no. Yes, because you will spend more time on organic chemistry than you would spend in a course on underwater basket weaving. And no, because those who say its so tough have studied inefficiently. Ask around, and you will find that most students think of organic chemistry as a memorization game. *This is not true!* Former organic chemistry students perpetuate the false rumor that organic chemistry is the toughest class on campus, because it makes them feel better about the poor grades that they received.

If it's not about memorizing, then what is it? To answer this question, let's compare organic chemistry to a movie. Picture in your mind a movie where the plot changes every second. The "Usual Suspects" is an excellent example. If you're in a movie theatre watching a movie like that, you can't leave even for a second because you would miss something important to the plot. So you try your hardest to wait until the movie is over before going to the bathroom. Sound familiar?

Organic chemistry is very much the same. It is one long story, and the story actually makes sense if you pay attention. The plot constantly develops, and everything ties into the plot. If your attention wanders for too long, you could easily get lost.

OK, so it's a long movie. But don't I need to memorize it? Of course, there are some things you need to memorize. You need to know some important terminology and some other concepts that require a bit of memorization, but the amount of pure memorization is not that large. If I were to give you a list of 100 numbers, and I asked you to memorize them all for an exam, you would probably be very upset by this. But at the same time, you can probably tell me at least 10 telephone numbers off the top of your head. Each one of those has 10 digits (including the area codes). You never sat down to memorize all 10 telephone numbers. Rather, over time you slowly became accustomed to dialing those numbers until the point that you knew them. Let's see how this works in our movie analogy.

You probably know at least one person who has seen one movie more than five times and can quote every line by heart. How can this person do that? It's *not* because he or she tried to memorize the movie. The first time you watch a movie, you learn the plot. After the second time, you understand why individual scenes are necessary to develop the plot. After the third time, you understand why the dialogue was necessary to develop each scene. After the fourth time, you are quoting many of the lines by heart. *Never at any time did you make an effort to memorize the lines.* You know them *because they make sense* in the grand scheme of the plot. If I were to give you a screenplay for a movie and ask you to memorize as much as you can in 10 hours, you would probably not get very far into it. If, instead, I put you in a room for 10 hours and played the same movie over again five times, you would know most of the movie by heart, without even trying. You would know everyone's names, the order of the scenes, much of the dialogue, and so on.

Organic chemistry is exactly the same. It's not about memorization. It's all about making sense of the plot, the scenes, and the individual concepts that make up our story. Of course you will need to remember all of the terminology, but with enough practice, the terminology will become second nature to you. So here's a brief preview of the plot.

THE PLOT

The first half of our story builds up to reactions, and we learn about the characteristics of molecules that help us understand reactions. We begin by looking at atoms, the building blocks of molecules, and what happens when they combine to form bonds. We focus on special bonds between certain atoms, and we see how the nature of bonds can affect the shape and stability of molecules. At this point, we need a vocabulary to start talking about molecules, so we learn how to draw and name molecules. We see how molecules move around in space, and we explore the relationships between similar types of molecules. At this point, we know the important characteristics of molecules, and we are ready to use our knowledge to explore reactions.

Reactions take up the rest of the course, and they are typically broken down into chapters based on functional groups. Within each of these chapters, there is actually a subplot that fits into the grand story.

HOW TO USE THIS BOOK

This book will help you study more efficiently so that you can avoid wasting countless hours. It will point out the major scenes in the plot of organic chemistry. The book will review the critical principles and explain why they are relevant to the rest of the course. In each section, you will be given the tools to better understand your textbook and lectures. In other words, you will learn the language of organic chemistry. *This book cannot replace your textbook, your lectures, or other forms of studying.* This book is not the Cliff Notes of Organic Chemistry. It focuses on the basic concepts that will empower you to do well if you go to lectures and study in addition to using this book. To best use this book, you need to know how to study in this course.

HOW TO STUDY

There are two separate aspects to this course:

   1. Understanding principles

   2. Solving problems

Although these two aspects are completely different, instructors will typically gauge your understanding of the principles by testing your ability to solve problems. So you must master both aspects of the course. The principles are in your textbook and in your lecture notes, but *you* must discover how to solve problems. Most students have a difficult time with this task. In this book, we explore some step-by-step processes for analyzing problems. There is a very simple habit that you must form immediately: *learn to ask the right questions.*

   If you go to a doctor with a pain in your stomach, you will get a series of questions: How long have you had the pain? Where is the pain? Does it come and go, or is it constant? What was the last thing you ate? and so on. The doctor is doing two very important and very different things. First, he has learned the right questions to ask. Next, he applies the knowledge he has together with the information he has gleaned to arrive at the proper diagnosis. Notice that the first step is asking the right questions.

   Let's imagine that you want to sue McDonald's because you spilled hot coffee in your lap. You go to an attorney and she asks you a series of questions that enable her to apply her knowledge to your case. Once again, the first step is asking questions.

   In fact, in any profession or trade, the first step of diagnosing a problem is always to ask questions. Let's say you are trying to decide if you really want to be a doctor. There are some tough, penetrating questions that you should be asking yourself. It all boils down to learning how to ask the right questions.

   The same is true with solving problems in this course. Unfortunately, you are expected to learn how to do this on your own. In this book, we will look at some common types of problems and we will see what questions you should be asking in those circumstances. More importantly, we will also be developing skills that will allow you to figure out what questions you should be asking for a problem that you have never seen before.

   Many students freak out on exams when they see a problem that they can't do. If you could hear what was going on in their minds, it would sound something like this: "I can't do it . . . I'm gonna flunk." These thoughts are counterproductive and a waste of precious time. Remember that when all else fails, there is always one question that you can ask yourself: "What questions should I be asking right now?"

   The only way to truly master problem-solving is to practice problems every day, consistently. You will never learn how to solve problems by just reading a book. You must try, and fail, and try again. You must learn from your mistakes. You must get frustrated when you can't solve a problem. That's the learning process.

   The worst thing you can do is to read through the solutions manual and think that you now know how to solve problems. It doesn't work that way. If you want an

A, you will need to sweat a little (no pain, no gain). And that doesn't mean that you should spend day and night memorizing. Students who focus on memorizing will experience the pain, but few of them will get an A.

The simple formula: Review the principles until you understand how each of them fits into the plot; then *focus all of your remaining time on solving problems.* Don't worry. The course is not that bad if you approach it with the right attitude. This book will act as a road map for your studying efforts.

CONTENTS

CHAPTER 1

BOND-LINE DRAWINGS

To do well in organic chemistry, you must first learn to interpret the drawings that organic chemists use. When you see a drawing of a molecule, it is absolutely critical that you can read all of the information contained in that drawing. Without this skill, it will be impossible to master even the most basic reactions and concepts.

Molecules can be drawn in many ways:

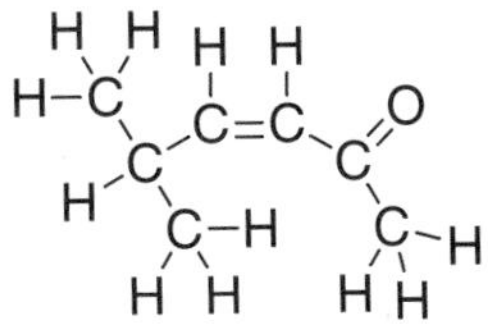

$(CH_3)_2CH=CHCOCH_3$

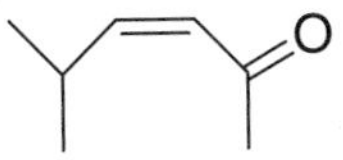

Without a doubt, the last structure (bond-line drawing) is the quickest to draw, the quickest to read, and the best way to communicate. Open your textbook to any page in the second half and you will find that every page is plastered with bond-line drawings. Most students will gain a familiarity with these drawings over time, not realizing how absolutely critical it is to be able to read these drawings fluently. This chapter will help you develop your skills in reading these drawings quickly and fluently.

1.1 HOW TO READ BOND-LINE DRAWINGS

Bond-line drawings show the carbon skeleton (the connections of all the carbon atoms that build up the backbone, or skeleton, of the molecule) with any functional groups that are attached, such as $-OH$ or $-Br$. Lines are drawn in a zigzag format, so that the end of every line represents a carbon atom. For example, the following compound has 6 carbon atoms:

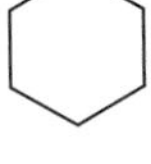

It is a common mistake to forget that the ends of lines represent carbon atoms as well. For example, the following molecule has six carbon atoms (make sure you can count them):

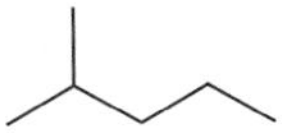

Double bonds are shown with two lines, and triple bonds are shown with three lines:

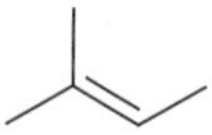

When drawing triple bonds, be sure to draw them in a straight line rather than zigzag, because triple bonds are linear (there will be more about this in the chapter on geometry). This can be quite confusing at first, because it can get hard to see just how many carbon atoms are in a triple bond, so let's make it clear:

is the same as 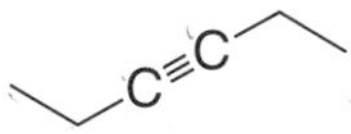   so this compound has
                        6 carbon atoms

Don't let triple bonds confuse you. The two carbon atoms of the triple bond and the two carbons connected to them are drawn in a straight line. All other bonds are drawn as a zigzag:

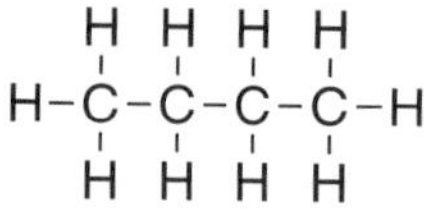

H H H H
| | | |
H-C-C-C-C-H   is drawn like this:
| | | |
H H H H

BUT

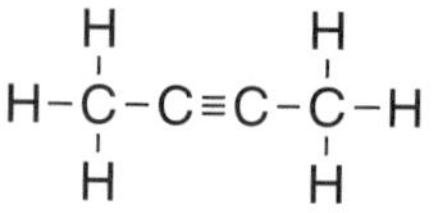

H   H
| |
H-C-C≡C-C-H   is drawn like this:
| |
H   H

EXERCISE 1.1 Count the number of carbon atoms in each of the following drawings:

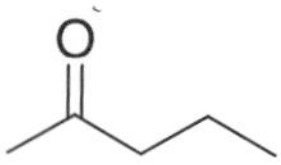

Answer The first compound has six carbon atoms, and the second compound has five carbon atoms:

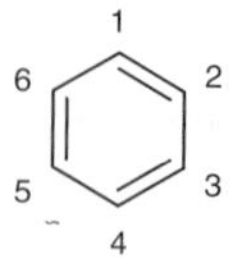

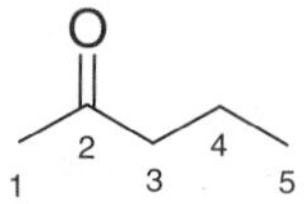

PROBLEMS Count the number of carbon atoms in each of the following drawings.

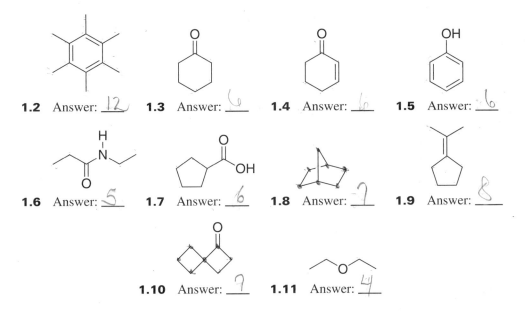

1.2 Answer: 12   **1.3** Answer: 6   **1.4** Answer: 6   **1.5** Answer: 6

1.6 Answer: 5   **1.7** Answer: 6   **1.8** Answer: 7   **1.9** Answer: 8

1.10 Answer: 7   **1.11** Answer: 4

Now that we know how to count carbon atoms, we must learn how to count the hydrogen atoms in a bond-line drawing of a molecule. The hydrogen atoms are not shown, and this is why it is so easy and fast to draw bond-line drawings. Here is the rule for determining how many hydrogen atoms there are on each carbon atom: *neutral carbon atoms always have a total of four bonds*. In the following drawing, the highlighted carbon atom is showing only two bonds:

We see only two bonds
connected to this carbon atom

Therefore, it is assumed that there are two more bonds to hydrogen atoms (to give a total of four bonds). This is what allows us to avoid drawing the hydrogen atoms and to save so much time when drawing molecules. It is assumed that the average person knows how to count to four, and therefore is capable of determining the number of hydrogen atoms even though they are not shown.

So you only need to count the number of bonds that you can see on a carbon atom, and then you know that there should be enough hydrogen atoms to give a total of four bonds to the carbon atom. After doing this many times, you will get to a point where you do not need to count anymore. You will simply get accustomed to seeing these types of drawings, and you will be able to instantly "see" all of the hydrogen atoms without counting them. Now we will do some exercises that will help you get to that point.

EXERCISE 1.12   The following molecule has 14 carbon atoms. Count the number of hydrogen atoms connected to each carbon atom.

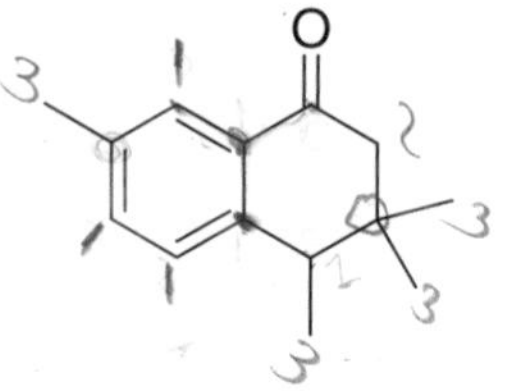

Answer:

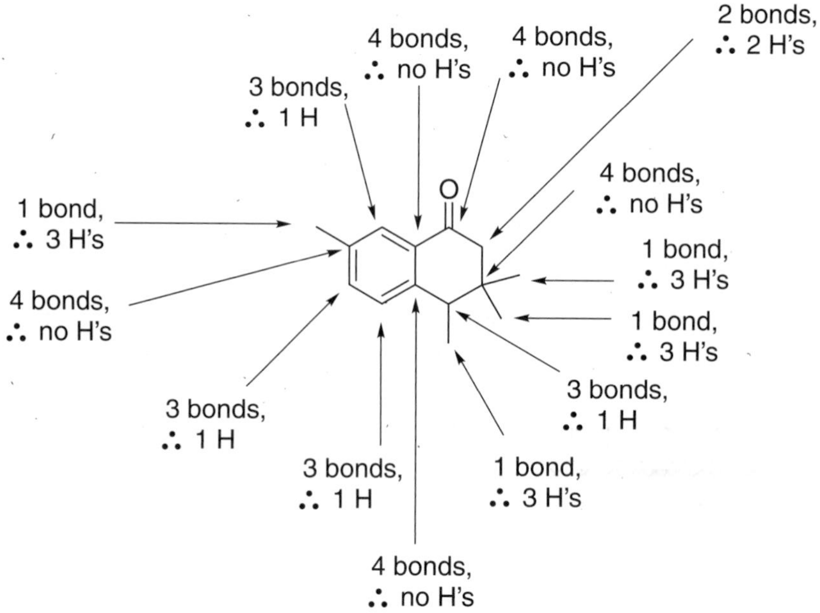

PROBLEMS   For each of the following molecules, count the number of hydrogen atoms connected to each carbon atom. The first problem has been solved for you (the numbers indicate how many hydrogen atoms are attached to each carbon).

1.13   **1.14** 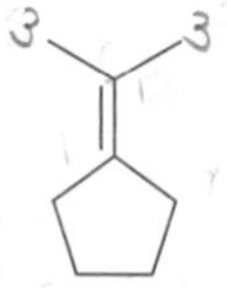   **1.15** 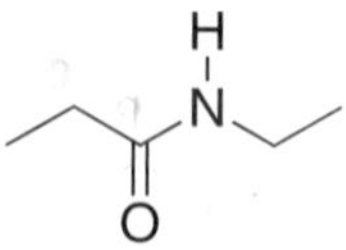   **1.16**

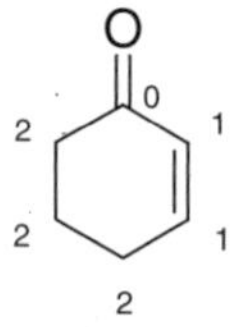

1.17 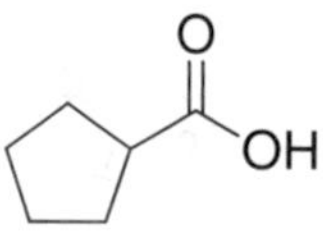   **1.18**    **1.19**    **1.20**

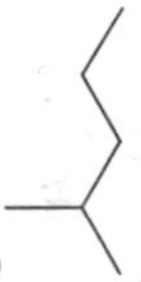

Now we can understand why we save so much time by using bond-line drawings. Of course, we save time by not drawing every C and H. But, there is an even larger benefit to using these drawings. Not only are they easier to draw, but they are easier to read as well. Take the following reaction for example:

$$(CH_3)_2CH{=}CHCOCH_3 \quad \xrightarrow[\text{Pt}]{\text{H}_2} \quad (CH_3)_2CH_2CH_2COCH_3$$

It is somewhat difficult to see what is happening in the reaction. You need to stare at it for a while to see the change that took place. However, when we redraw the reaction using bond-line drawings, the reaction becomes very easy to read immediately:

As soon as you see the reaction, you immediately know what is happening. In this reaction we are converting a double bond into a single bond by adding two hydrogen atoms across the double bond. Once you get comfortable reading these drawings, you will be better equipped to see the changes taking place in reactions.

1.2 HOW TO DRAW BOND-LINE DRAWINGS

Now that we know how to read these drawings, we need to learn how to draw them. Take the following molecule as an example:

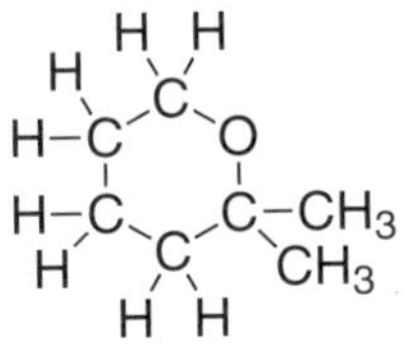

To draw this as a bond-line drawing, we focus on the carbon skeleton, making sure to draw any atoms other than C and H. All atoms other than carbon and hydrogen *must* be drawn. So the example above would look like this:

A few pointers may be helpful before you do some problems.

  1. Don't forget that carbon atoms in a straight chain are drawn in a zigzag format:

  is drawn like this:

  2. When drawing double bonds, try to draw the other bonds as far away from the double bond as possible:

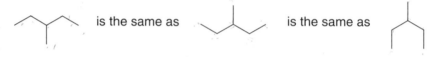

  is much better than

  BAD

  3. When drawing zigzags, it does not matter in which direction you start drawing:

  is the same as   is the same as

PROBLEMS For each structure below, draw the bond-line drawing in the box provided.

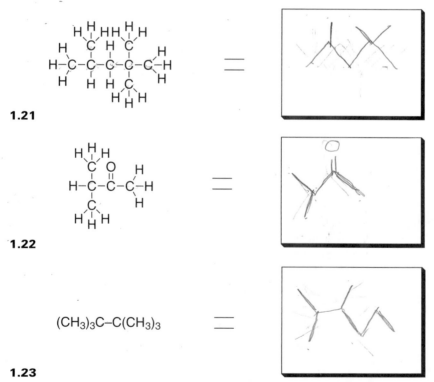

1.21

1.22

$(CH_3)_3C-C(CH_3)_3$

1.23

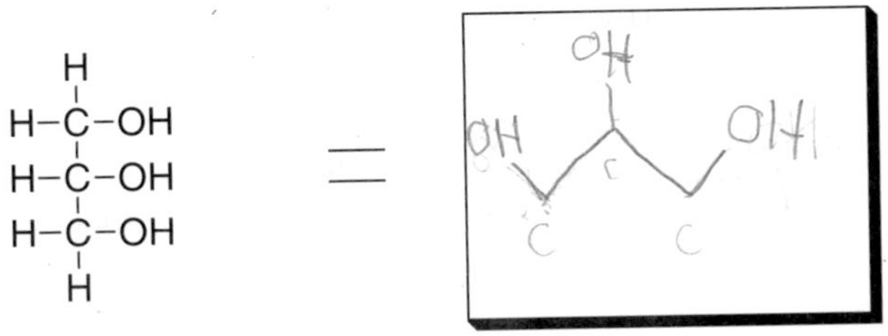

H
|
H–C–OH
|
H–C–OH $\equiv$
|
H–C–OH
|
H

1.24

1.3 MISTAKES TO AVOID

1. *Never* draw a carbon atom with more than four bonds. This is a big no-no. Carbon atoms only have four orbitals; therefore, carbon atoms can form only four bonds (bonds are formed when orbitals of one atom overlap with orbitals of another atom). This is true of all second-row elements, and we discuss this in more detail in the chapter on drawing resonance structures.

2. When drawing a molecule, you should either show all of the H's and all of the C's, or draw a bond-line drawing where the C's and H's are not drawn. You *cannot* draw the C's without also drawing the H's:

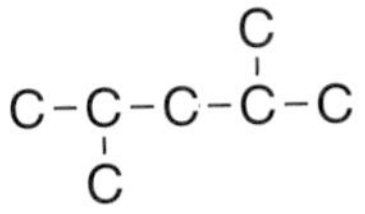

C
|
C–C–C–C–C   Never do this
|
C

This drawing is no good. Either leave out the C's (which is preferable) or put in the H's:

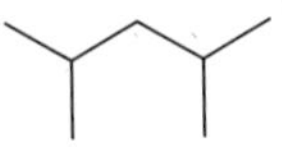

or

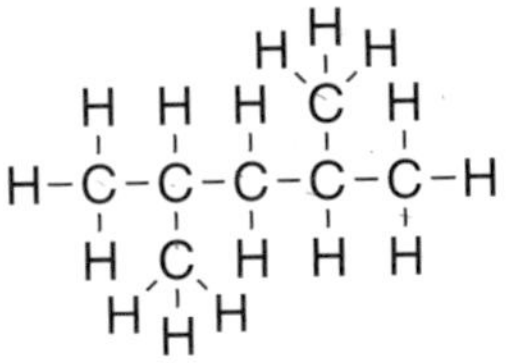

H H H H H
| | | | |
H–C–C–C–C–C–H
| | | | |
H C H H H
H H H

3. When drawing each carbon atom in a zigzag, try to draw all of the bonds as far apart as possible:

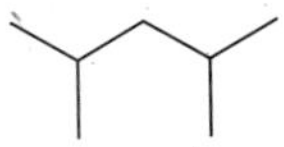

is better than

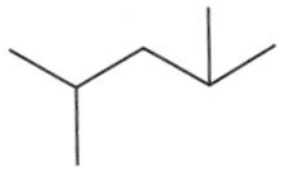

4. In bond-line drawings, we do draw any H's that are connected to atoms other than carbon. For example,

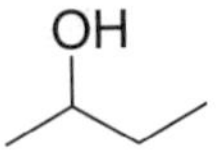

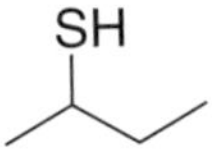

OH   SH

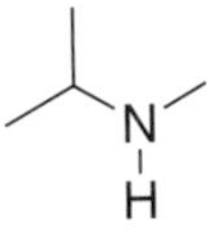

1.4 MORE EXERCISES

First, open your textbook and flip through the pages in the second half. Choose any bond-line drawing and make sure that you can say with confidence how many carbon atoms you see and how many hydrogen atoms are attached to each of those carbon atoms.

Now try to look at the following reaction and determine what changes took place:

Do not worry about *how* the changes took place. You will understand that later when you learn the mechanism of the reaction. For now, just focus on explaining what change took place. For the example above, we can say that we *added* two hydrogen atoms to the molecule (one on either end of the double bond).

Consider another example:

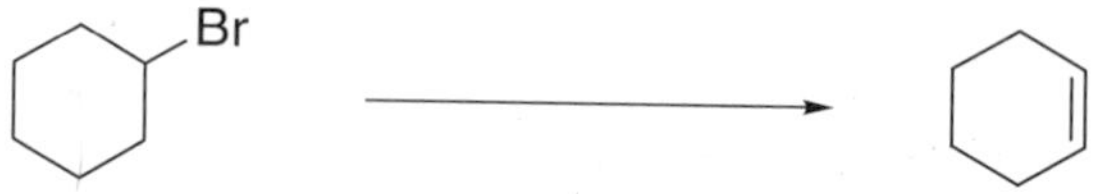

In this example, we have *eliminated* an H and a Br to form a double bond. (We will see later that it is actually H^+ and Br^- that are eliminated, when we get into the chapters on mechanisms). If you cannot see that an H was eliminated, then you will need to count the number of hydrogen atoms in the starting material and compare it with the product:

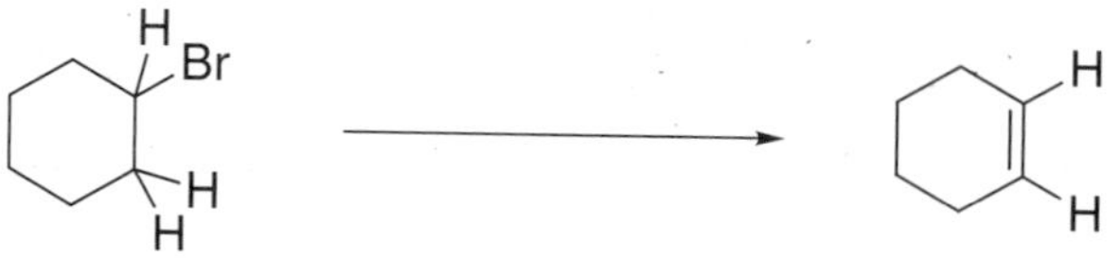

Now consider one more example:

In this example, we have *substituted* a bromine with a chlorine.

PROBLEMS For each of the following reactions, clearly state what change has taken place. In each case your sentence should start with one of the following opening clauses: we have added . . . , we have eliminated . . . , or we have substituted . . . .

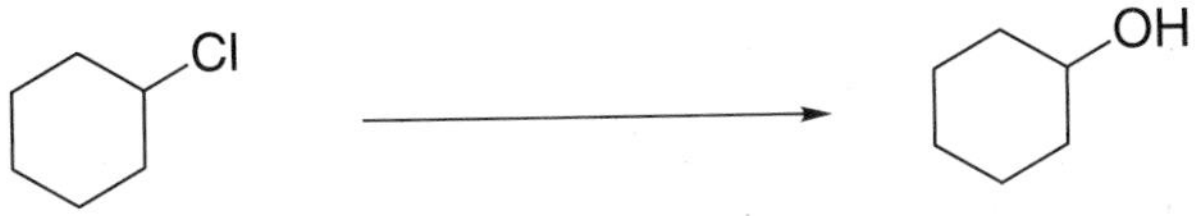

1.25

Answer: _we have substituted a chlorine with a hydroxide_

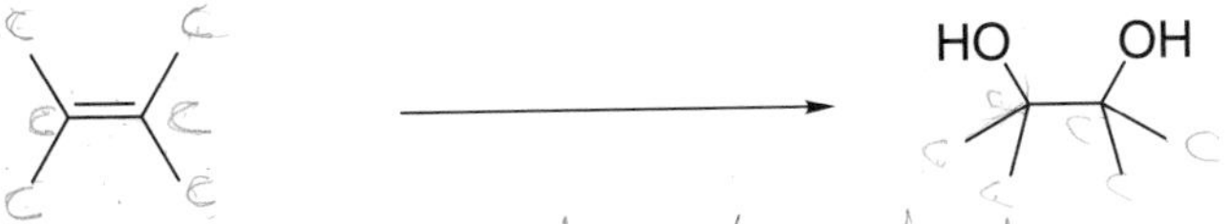

1.26

Answer: _we have added two hydrogen atoms_   X
we have substituted two hydroxide atoms

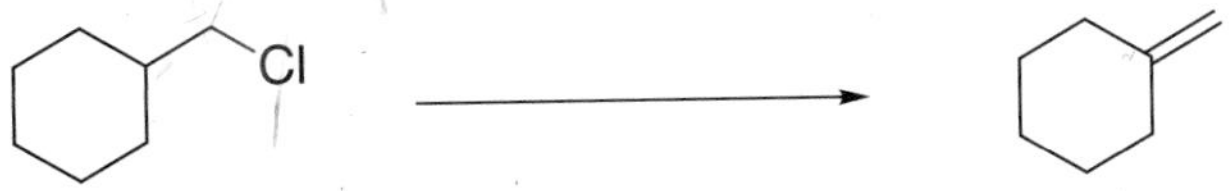

1.27

Answer: _We have eliminated a chlorine atom and to form a double bond, hydrogen_

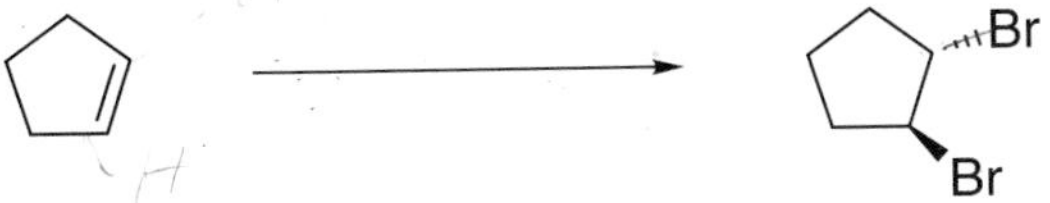

1.28

Answer: _we have added two bromine atoms to add 2 bromine atoms across a double bond_

1.29

Answer: _we have added two hydrogen atom_   X

1.30

Answer: _we have substituted I atom with a sulfur atom SH_

1.31

Answer: _We have added two hydrogen atoms_

1.32

Answer: _We have eliminated two hydrogen atoms_

1.5   IDENTIFYING FORMAL CHARGES

Formal charges are charges (either positive or negative) that we must often include on our drawings. They are extremely important. If you don't draw a formal charge when it is supposed to be drawn, then your drawing will be incomplete (and wrong). So you must learn how to identify when you need formal charges and how to draw them. If you cannot do this, then you will not be able to draw resonance structures (which we see in the next chapter), and if you can't do that, then you will have a very hard time passing this course.

To understand what formal charges are, we begin by learning how to calculate formal charges. By doing this, you will understand what formal charges are. So how do we calculate formal charges?

When calculating the formal charge on an atom, we first need to know the number of valence electrons the atom is *supposed* to have. We can get this number from the periodic table. The column of the periodic table that the atom is in will tell us how many valence electrons there are (valence electrons are the electrons in the valence shell, or the outermost shell of electrons—you probably remember this from high school chemistry). For example, carbon is in the fourth column, and therefore has four valence electrons. Now you know how to determine how many electrons the atom is supposed to have.

Next we look in our drawing and ask how many electrons the atom *actually has* in the drawing. But how do we count this?

Let's see an example. Consider the central carbon atom in the compound below:

$$H_3C-\overset{\overset{\displaystyle \ddot{O}-H}{|}}{\underset{\underset{\displaystyle H}{|}}{C}}-CH_3$$

Remember that every bond represents two electrons being shared between two atoms. Begin by splitting each bond apart, placing one electron on this atom and one electron on that atom:

Now count the number of electrons immediately surrounding the central carbon atom:

There are four electrons. This is the number of electrons that the atom actually has.

Now we are in a position to compare how many valence electrons the atom is *supposed* to have (in this case, four) with how many valence electrons it *actually* has (in this case, four). Since these numbers are the same, the carbon atom has no formal charge. This will be the case for most of the atoms in the structures you will draw in this course. But in some cases, the number of electrons the atom is supposed to have and the number of electrons the atom actually has will be different. In those cases, there will be a formal charge. So let's see an example of an atom that has a formal charge.

Consider the oxygen atom in the structure below:

Let's begin by asking how many valence electrons oxygen atoms are *supposed* to have. Oxygen is in the sixth column of the periodic table, so oxygen should have six valence electrons. Next, we need to look at the oxygen atom in this compound and ask how many valence electrons it *actually* has. So, we redraw the molecule by splitting up the C–O bond:

In addition to the electron on the oxygen from the C–O bond, the oxygen also has three lone pairs. A lone pair is when you have two electrons that are not being used to form a bond. Lone pairs are drawn as two dots on an atom, and the oxygen above has three of these lone pairs. You must remember to count each lone pair as two electrons. So we see that the oxygen actually has seven electrons, which is one more electron than it is supposed to have. Therefore, it will have a negative charge:

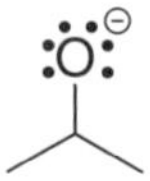

EXERCISE 1.33 Consider the nitrogen atom in the compound below and determine if it has a formal charge:

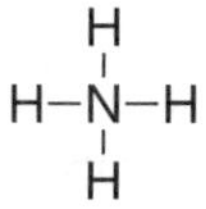

Answer Nitrogen is in the fifth column of the periodic table so it should have five electrons. Now we count how many it actually has:

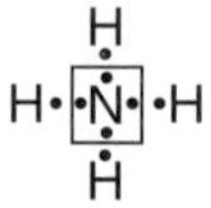

It only has four. So, it has one less electron than it is supposed to have. Therefore, this nitrogen atom has a positive charge:

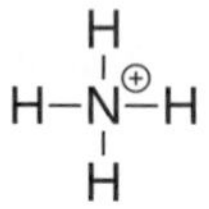

PROBLEMS For each of the compounds below determine if the oxygen or nitrogen atom in the molecule has a formal charge. If there is a charge, draw the charge on the structure.

1.34 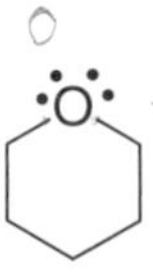   1.35    1.36 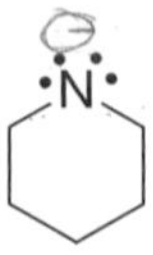   1.37

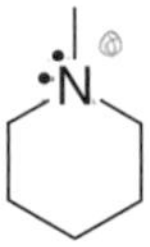

1.38 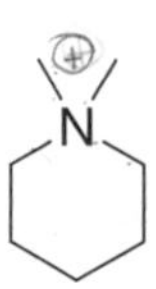   1.39 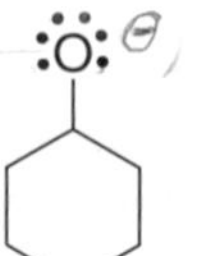   1.40 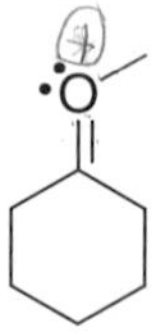   1.41

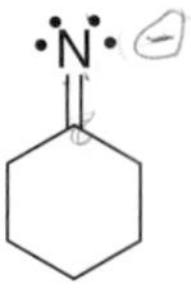

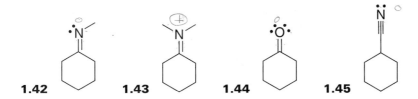

1.42   **1.43**   **1.44**   **1.45**

This brings us to the most important atom of all: carbon. We saw before that carbon always has four bonds. This allows us to ignore the hydrogen atoms when drawing bond-line structures, because it is assumed that we know how to count to four and can figure out how many hydrogen atoms are there. When we said that, we were only talking about carbon atoms without formal charges (most carbon atoms in most structures will not have formal charges). But now that we have learned what a formal charge is, let's consider what happens when carbon has a formal charge.

If carbon bears a formal charge, then we cannot just assume the carbon has four bonds. In fact, it will have only three. Let's see why. Let's first consider C^+, and then we will move on to C^-.

If carbon has a *positive* formal charge, then it has only three electrons (it is supposed to have four electrons, because carbon is in the fourth column of the periodic table). Since it has only three electrons, it can form only three bonds. That's it. So, a carbon with a positive formal charge will have only three bonds, and you should count hydrogen atoms with this in mind:

No hydrogen atoms on this C^+ 1 hydrogen atom on this C^+ 2 hydrogen atoms on this C^+

Now let's consider what happens when we have a carbon with a *negative* formal charge. The reason it has a negative formal charge is because it has one more electron than it is supposed to have. Therefore, it has five electrons. Two of these electrons form a lone pair, and the other three electrons are used to form bonds:

$$H-\overset{\overset{\displaystyle H}{|}}{\underset{\underset{\displaystyle H}{|}}{C}}:\overset{\ominus}{}$$

We have the lone pair, because we can't use each of the five electrons to form a bond. Carbon can *never* have five bonds. Why not? Electrons exist in regions of space called orbitals. These orbitals can overlap with orbitals from other atoms to form bonds, or the orbitals can contain two electrons (which is called a lone pair). Carbon has only four orbitals, so there is no way it could possibly form five bonds—it does not have five orbitals to use to form those bonds. This is why a carbon atom with a negative charge will have a lone pair (if you look at the drawing above, you will count four orbitals—one for the lone pair and then three more for the bonds).

Therefore, a carbon atom with a negative charge can also form only three bonds (just like a carbon with a positive charge). When you count hydrogen atoms, you should keep this in mind:

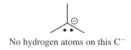

No hydrogen atoms on this C⁻

1.6 FINDING LONE PAIRS THAT ARE NOT DRAWN

From all of the cases above (oxygen, nitrogen, carbon), you can see why you have to know how many lone pairs there are to figure out the formal charge on an atom. Similarly, you have to know the formal charge to figure out how many lone pairs there are on an atom. Take the case below with the nitrogen atom shown:

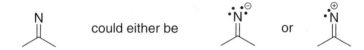

If the lone pairs were drawn, then we would be able to figure out the charge (two lone pairs would mean a negative charge and one lone pair would mean a positive charge). Similarly, if the formal charge was drawn, we would be able to figure out how many lone pairs there are (a negative charge would mean two lone pairs and a positive charge would mean one lone pair).

So you can see that drawings must include either lone pairs or formal charges. The convention is to always show formal charges and to leave out the lone pairs. This is much easier to draw, because you usually won't have more than one charge on a drawing (if even that), so you get to save time by not drawing every lone pair on every atom.

Now that we have established that formal charges must *always* be drawn and that lone pairs are usually *not* drawn, we need to get practice in how to see the lone pairs when they are not drawn. This is not much different from training yourself to see all the hydrogen atoms in a bond-line drawing even though they are not drawn. If you know how to count, then you should be able to figure out how many lone pairs are on an atom where the lone pairs are not drawn.

Let's see an example to demonstrate how you do this:

In this case, we are looking at an oxygen atom. Oxygen is in the sixth column of the periodic table, so it is supposed to have six electrons. Then, we need to take the formal charge into account. This oxygen atom has a negative charge, which means one extra electron. Therefore, this oxygen atom must have $6 + 1 = 7$ electrons. Now we can figure out how many lone pairs there are.

The oxygen has one bond, which means that it is using one of its seven electrons to form a bond. The other six must be in lone pairs. Since each lone pair is two electrons, this must mean that there are three lone pairs:

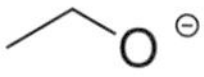

is the same as

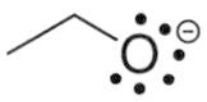

Let's review the process:

1. Count the number of electrons the atom should have according to the periodic table.
2. Take the formal charge into account. A negative charge means one more electron, and a positive charge means one less electron.
3. Now you know the number of electrons the atom actually has. Use this number to figure out how many lone pairs there are.

Now we need to get used to the common examples. Although it is important that you know how to count and determine numbers of lone pairs, it is actually much more important to get to a point where you don't have to waste time counting. You need to get familiar with the common situations you will encounter. Let's go through them methodically.

When oxygen has no formal charge, it will have two bonds and two lone pairs:

is the same as

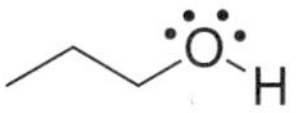

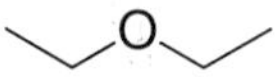

is the same as

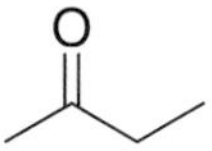

is the same as

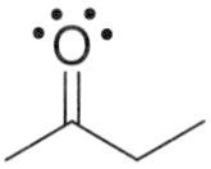

If oxygen has a negative formal charge, then it must have one bond and three lone pairs:

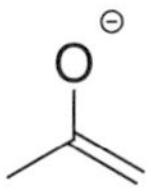

is the same as

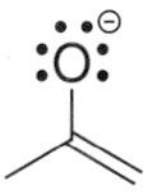

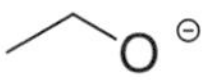

is the same as

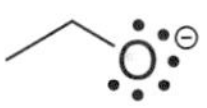

If oxygen has a positive charge, then it must have three bonds and one lone pair:

$\overset{\oplus}{O}H_2$ is the same as   (structure with H, O⁺, H)

(structure with H, O⁺)   is the same as   (structure)

(structure with O⁺, H)   is the same as   (structure)

EXERCISE 1.46   Draw in all lone pairs in the following structure:

Answer   The oxygen has a positive formal charge and three bonds. You should try to get to a point where you recognize that this must mean that the oxygen has one lone pair:

Until you get to the point where you can recognize this, you should be able to figure out the answer by counting.

Oxygen is supposed to have six electrons. This oxygen atom has a positive charge, which means it is missing an electron. Therefore, this oxygen atom must have $6 - 1 = 5$ electrons. Now, we can figure out how many lone pairs there are.

The oxygen has three bonds, which means that it is using three of its five electrons to form bonds. The other two must be in a lone pair. So there is only one lone pair.

PROBLEMS   Review the common situations above, and then come back to these problems. For each of the following structures, draw in all lone pairs. Try to recognize how many lone pairs there are *without* having to count. Then count to see if you were right.

1.47   **1.48**   **1.49**

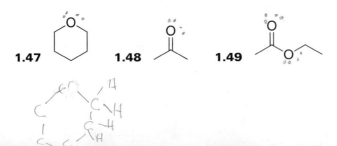

1.50 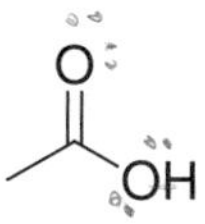   **1.51** 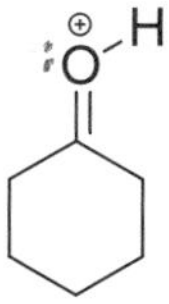   **1.52**

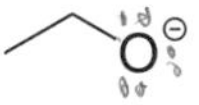

Now let's look at the common situations for nitrogen atoms. When nitrogen has no formal charge, it will have three bonds and one lone pair:

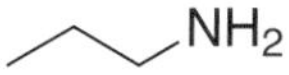

is the same as

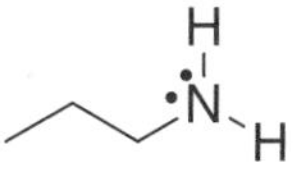

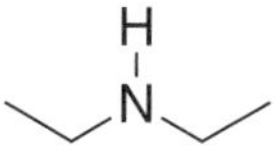

is the same as

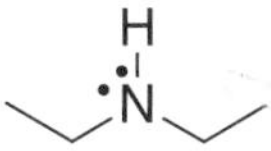

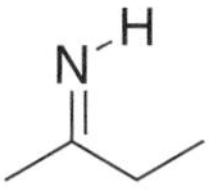

is the same as

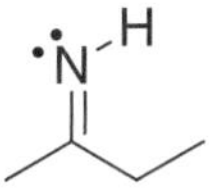

If nitrogen has a negative formal charge, then it must have two bonds and two lone pairs:

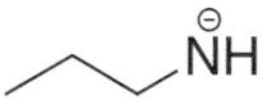

is the same as

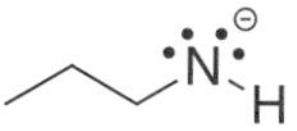

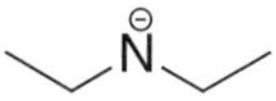

is the same as

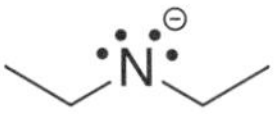

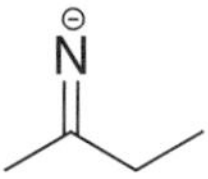

is the same as

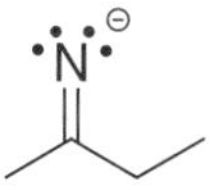

If nitrogen has a positive charge, then it must have four bonds and no lone pairs:

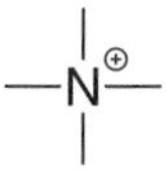

has no lone pairs

has no lone pairs

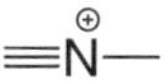

has no lone pairs

EXERCISE 1.53 Draw all lone pairs in the following structure:

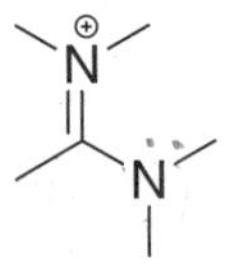

Answer The top nitrogen has a positive formal charge and four bonds. You should try to get to a point where you recognize that this must mean that this nitrogen has no lone pairs. The bottom nitrogen has no formal charge and three bonds. You should try to get to a point where you recognize that this must mean that this nitrogen has one lone pair:

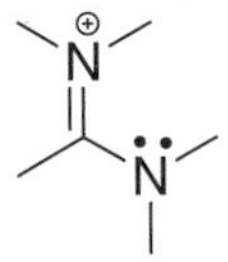

Until you get to the point where you can recognize this, you should be able to figure out the answer by counting. Nitrogen is supposed to have five electrons. The top nitrogen atom has a positive charge, which means it is missing an electron. This means that this nitrogen atom must have $5 - 1 = 4$ electrons. Now we can figure out how many lone pairs there are. Since this nitrogen has four bonds, it is using all of its electrons to form bonds. So there is no lone pair on this nitrogen atom.

The bottom nitrogen atom has no formal charge, so this nitrogen atom has five electrons. It has three bonds, which means that there are two electrons left over, and they form a lone pair.

PROBLEMS Review the common situations for nitrogen, and then come back to these problems. For each of the following structures, draw in all lone pairs. Try to recognize how many lone pairs there are *without* having to count. Then count to see if you were right.

1.54

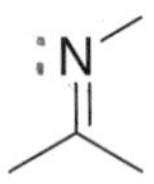

1.55

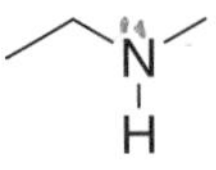

1.56

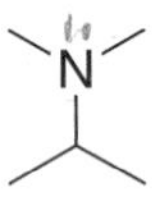

1.57

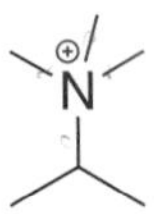

1.58

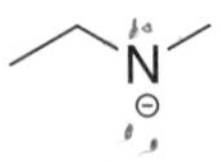

1.59

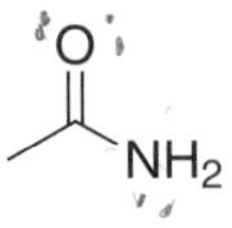

MORE PROBLEMS For each of the following structures, draw in all lone pairs (remember from the previous section that C^+ has no lone pairs and C^- has one lone pair).

1.60

1.61

1.62

1.63

1.64

1.65

1.66 $^{\ominus}O-C\equiv N$

1.67 $O=C=N^{\ominus}$

1.68

CHAPTER 2

RESONANCE

In this chapter, you will learn the tools that you need to draw resonance structures with proficiency. I cannot adequately stress the importance of this skill. Resonance is the one topic that permeates the entire subject matter from start to finish. It finds its way into every chapter, into every reaction, and into your nightmares if you do not master the rules of resonance. You cannot get an A in this class without mastering resonance. So what is resonance? And why do we need it?

2.1 WHAT IS RESONANCE?

In Chapter 1, we introduced one of the best ways of drawing molecules, bond-line structures. They are fast to draw and easy to read, but they have one major deficiency: they do not describe molecules perfectly. In fact, no drawing method can completely describe a molecule using only a single drawing. Here is the problem.

Although our drawings are very good at showing which atoms are connected to each other, our drawings are not good at showing where all of the electrons are, because electrons aren't really solid particles that can be in one place at one time. All of our drawing methods treat electrons as particles that can be placed in specific locations. Instead, it is best to think of electrons as *clouds of electron density*. We don't mean that electrons fly around in clouds; we mean that electrons *are* clouds. These clouds often spread themselves across large regions of a molecule.

So how do we represent molecules if we can't draw where the electrons are? The answer is resonance. We use the term *resonance* to describe our solution to the problem: we use more than one drawing to represent a single molecule. We draw several drawings, and we call these drawings *resonance structures*. We meld these drawings into one image in our minds. To better understand how this works, consider the following analogy.

Your friend asks you to describe what a nectarine looks like, because he has never seen one. You aren't a very good artist so you say the following:

> *Picture a peach in your mind, and now picture a plum in your mind. Well, a nectarine has features of both: the inside tastes like a peach, but the outside is smooth like a plum. So take your image of a peach together with your image of a plum and meld them together in your mind into one image. That's a nectarine.*

It is important to realize that a nectarine does not switch back and forth every second from being a peach to being a plum. A nectarine is a nectarine all of the time.

The image of a peach is not adequate to describe a nectarine. Neither is the image of a plum. But by imagining both together at the same time, you can get a sense of what a nectarine looks like.

The problem with drawing molecules is similar to the problem above with the nectarine. No single drawing adequately describes the nature of the electron density spread out over the molecule. To solve this problem, we draw several drawings and then meld them together in our mind into one image. Just like the nectarine.

Let's see an example:

The compound above has two important resonance structures. Notice that we separate resonance structures with a straight, two-headed arrow, and we place brackets around the structures. The arrow and brackets indicate that they are resonance structures *of one molecule.* The molecule is not flipping back and forth between the different resonance structures. The electrons in the molecule are not actually moving at all.

Now that we know why we need resonance, we can begin to understand why resonance structures are so important. Ninety-five percent of the reactions that you will see in this course occur because one molecule has a region of low electron density and the other molecule has a region of high electron density. They attract each other in space, which causes a reaction. So, to predict how and when two molecules will react with each other, we need first to predict where there is low electron density and where there is high electron density. We need to have a firm grasp of resonance to do this. In this chapter, we will see many examples of how to predict the regions of low or high electron density by applying the rules of drawing resonance structures.

2.2 CURVED ARROWS: THE TOOLS FOR DRAWING RESONANCE STRUCTURES

In the beginning of the course, you might encounter problems like this: here is a drawing; now draw the other resonance structures. But later on in the course, it will be assumed and expected that you can draw all of the resonance structures of a compound. If you cannot actually do this, you will be in big trouble later on in the course. So how do you draw all of the resonance structures of a compound? To do this, you need to learn the tools that help you: curved arrows.

Here is where it can be confusing as to what is exactly going on. These arrows do not represent an actual process (such as electrons moving). This is an important point, because you will learn later about curved arrows used in drawing reaction mechanisms. Those arrows look exactly the same, but they actually do refer to the flow of electron density. In contrast, curved arrows here are used only as tools to help us draw all resonance structures of a molecule. The electrons are not actually moving. It can be tricky because we will say things like: "this arrow shows the electrons coming from here and going to there." But we don't actually mean that the electrons are moving; they are *not* moving. Since each drawing treats the electrons as particles stuck in one place, we will need to "move" the electrons to get from one drawing to another. Arrows are the tools that we use to make sure that we know how to draw all resonance structures for a compound. So, let's look at the features of these important curved arrows.

Every curved arrow has a *head* and a *tail*. It is essential that the head and tail of every arrow be drawn in precisely the proper place. *The tail shows where the electrons are coming from, and the head shows where the electrons are going* (remember that the electrons aren't really going anywhere, but we treat them as if they were so we can make sure to draw all resonance structures):

Tail   Head

Therefore, there are only two things that you have to get right when drawing an arrow: the tail needs to be in the right place and the head needs to be in the right place. So we need to see rules about where you can and where you cannot draw arrows. But first we need to talk a little bit about electrons, since the arrows are describing the electrons.

Electrons exist in orbitals, which can hold a maximum of two electrons. So there are only three options for any orbital:

- 0 electrons in the orbital
- 1 electron in the orbital
- 2 electrons in the orbital

If there are no electrons in the orbital, then there's nothing to talk about (there are no electrons there). If you have one electron in the orbital, it can overlap with another electron in a nearby orbital (forming a *bond*). If two electrons occupy the orbital, they fill the orbital (called a *lone pair*). So we see that electrons can be found in only two places: in bonds or in lone pairs. Therefore, electrons can only come from either a bond or a lone pair. Similarly, electrons can only go to form either a bond or a lone pair.

Let's focus on tails of arrows first. Remember that the tail of an arrow indicates where the electrons are coming from. So the tail has to come from a place that has electrons: either from a bond or from a lone pair. Consider the following resonance structures as an example:

How do we get from the first structure to the second one? Notice that the electrons that make up the double bond have been "moved." This is an example of electrons coming from a bond. Let's see the arrow showing the electrons coming from the bond and going to form another bond:

Now let's see what it looks like when electrons come from a lone pair:

Never draw an arrow that comes from a positive charge. The tail of an arrow must come from a spot that has electrons.

Heads of arrows are just as simple as tails. The head of an arrow shows where the electrons are going. So the head of an arrow must either point directly in between two atoms to form a bond,

or it must point to an atom to form a lone pair.

Never draw the head of an arrow going off into space:

Bad arrow

Remember that the head of an arrow shows where the electrons are going. So the head of an arrow must point to a place where the electrons can go—either to form a bond or to form a lone pair.

2.3 THE TWO COMMANDMENTS

Now we know what curved arrows are, but how do we know when to push them and where to push them? First we need to learn where we *cannot* push arrows. There are two important rules that you can *never* violate when pushing arrows. They are the "two commandments" of drawing resonance structures:

    1. Thou shall not break a single bond.

    2. Thou shall not violate the octet rule.

Let's focus on one at a time.

    1. *Never break a single bond* when drawing resonance structures. By definition, resonance structures must have all the same atoms connected in the same order. Otherwise, they would be different compounds.

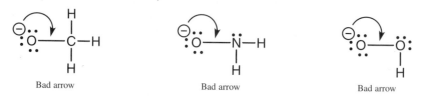

Never break a single bond

If you draw a tail of an arrow on a single bond, then you are breaking that single bond and violating the first commandment. So the first commandment is violated when a tail is not drawn in the right place.

    2. *Never violate the octet rule.* Let's review the octet rule. Atoms in the second row (C, N, O, F) have only four orbitals in their valence shell. Orbitals are used to form bonds and to hold lone pairs. Each bond is using one orbital and each lone pair is using one orbital. So the second-row elements can *never* have five or six bonds; the most is four. Similarly, they can never have four bonds and a lone pair, because this would also require five orbitals. For the same reason, they can never have three bonds and two lone pairs. Let's see some examples of arrow pushing that violates the octet rule:

Bad arrow   Bad arrow   Bad arrow

In each of these drawings, the central atom cannot form another bond because it does not have a fifth orbital that can be used. *This is impossible.* Don't ever do this.

The examples above are clear, but bond-line drawings are more difficult because we cannot see the hydrogen atoms (and, very often, we cannot see the lone pairs either; for now, we will continue to draw lone pairs to ease you into it). You have to train yourself to see the hydrogen atoms and to recognize when the octet rule is being violated:

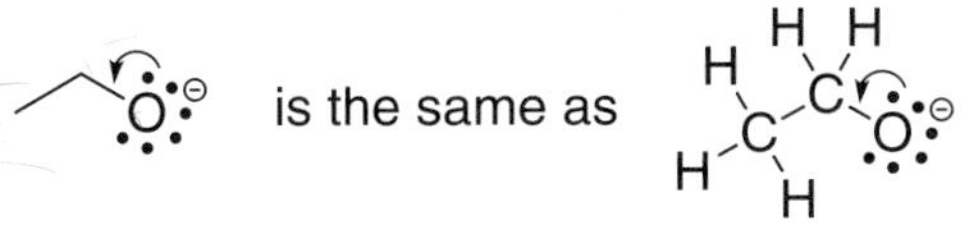

is the same as

At first it is difficult to see that the arrow on the left structure is violating the octet rule. But when we count the hydrogen atoms, we can see that the arrow above would give a carbon atom with five bonds.

If we draw a head of an arrow that forms a bond on an atom already using all four orbitals, then we are violating the second commandment. So the second commandment is violated when a head is not drawn in the right place.

Thus, the two commandments really do reflect the two parts of the arrow. A bad tail violates the first commandment and a bad head violates the second commandment.

EXERCISE 2.1 For the compound below, look at the arrow drawn on the structure and determine whether it violates either of the two commandments for drawing resonance structures:

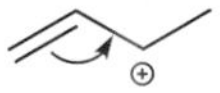

Answer First we need to ask if the first commandment has been violated: did we break a single bond? To determine this, we look at the *tail* of the arrow. If the tail of the arrow is coming from a single bond, then that means we are breaking that single bond. If the tail is coming from a double bond, then we have not violated the first commandment. In this example, the tail is on a double bond, so we did not violate the first commandment.

Now we need to ask if the second commandment has been violated: did we violate the octet rule? To determine this, we look at the *head* of the arrow. Are we forming a fifth bond? Remember that C^+ only has three bonds, not four. When we count the hydrogen atoms attached to this carbon, we see that there is only one hydrogen atom, not two, to give that carbon a total of three bonds. When we move the arrow shown above, the carbon will now get four bonds, and the second commandment has not been violated.

The arrow above is valid, because the two commandments were not violated.

PROBLEMS For each of the problems below, determine which arrows violate either one of the two commandments, and explain why. (Don't forget to count all hydrogen atoms and all lone pairs. You must do this to solve these problems.)

2.2 ___bad___

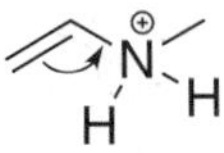

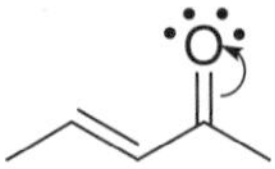

2.3 ___good___

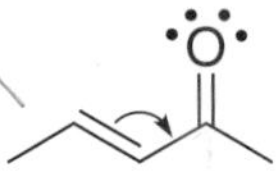

2.4 ___bad___

2.5 ___bad___

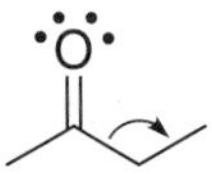

2.6 ___bad___

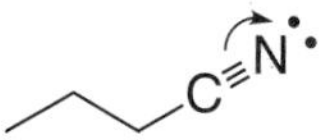

2.7 ___bad___

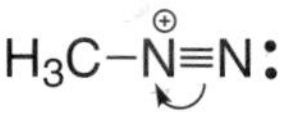

2.8 ___bad___

2.9 ___bad___

2.10 ___good___

2.11 $H_3C-\overset{\oplus}{N}\equiv N:$ ___good___

2.12 ___good bad___

2.4 DRAWING GOOD ARROWS

Now that we know how to identify good arrows and bad arrows, we need to get some practice drawing arrows. We know that the tail of an arrow must come either from a bond or a lone pair, and that the head of an arrow must go to form a bond or a lone pair. If we are given two resonance structures and are asked to show the arrow(s) that get us from one resonance structure to the other, it makes sense that we need to look for any bonds or lone pairs that are appearing or disappearing when going from one structure to another. Let's see this with an example.

Say we have two resonance structures:

How would we figure out what curved arrow to draw to get us from the drawing on the left to the drawing on the right? We must look at the difference between the two structures and ask, "How should we push the electrons to get from the first structure to the second structure?" Begin by looking for any double bonds or lone pairs that are disappearing. That will tell us where to put the tail of our arrow. In this example, there are no lone pairs disappearing, but there is a double bond disappearing. So we know that we need to put the tail of our arrow on the double bond.

Now, we need to know where to put the head of the arrow. We look for any lone pairs or double bonds that are appearing. We see that there is a new lone pair appearing on the oxygen. So now we know where to put the head of the arrow:

Notice that when we move a double bond up onto an atom to form a lone pair, it creates two formal charges: a positive charge on the carbon that lost its bond and a negative charge on the oxygen that got a lone pair. This is a very important issue. Formal charges were introduced in the last chapter, and now they will become instrumental in drawing resonance structures. For the moment, let's just focus on pushing arrows, and in the next section of this chapter, we will come back to focus on these formal charges.

It is pretty straightforward to see how to push only one arrow that gets us from one resonance structure to another. But what about when we need to push more than one arrow to get from one resonance structure to another? Let's do an example like that.

EXERCISE 2.13   For the two structures below, try to draw the curved arrows that get you from the drawing on the left to the drawing on the right:

Answer Let's analyze the difference between these two drawings. We begin by looking for any double bonds or lone pairs that are disappearing. We see that the oxygen is losing a lone pair, and the C=C on the bottom is also disappearing. This should automatically tell us that we need two arrows. To lose a lone pair and a double bond, we will need two tails.

Now let's look for any double bonds or lone pairs that are appearing. We see that a C=O is appearing and a C with a negative charge is appearing (remember that a C⁻ means a C with a lone pair). This tells us that we need two heads, which confirms that we need two arrows.

So we know we need two arrows. Let's start at the top. We lose a lone pair from the oxygen and form a C=O. Let's draw that arrow:

Notice that if we stopped here, we would be violating the second commandment. The central carbon atom is getting five bonds. To avoid this problem, we must immediately draw the second arrow. The C=C disappears (which solves our octet problem) and becomes a lone pair on the carbon. Now we can draw both arrows:

Arrow pushing is much like riding a bike. If you have never done it before, watching someone else will not make you an expert. You have to learn how to balance yourself. Watching someone else is a good start, but you have to get on the bike if you want to learn. You will probably fall a few times, but that's part of the learning process. The same is true with arrow pushing. The only way to learn is with practice.

Now it's time for you to get on the arrow-pushing bike. You would never be stupid enough to try riding a bike for the first time next to a steep cliff. Do not have your first arrow-pushing experience be during your exam. Practice right now!

PROBLEMS For each drawing, try to draw the curved arrows that get you from one drawing to the next. In many cases you will need to draw more than one arrow.

2.14

2.15

2.16

2.17

2.18

2.19

2.5 FORMAL CHARGES IN RESONANCE STRUCTURES

Now we know how to draw good arrows (and how to avoid drawing bad arrows). In the last section, we were given the resonance structures and just had to draw in the arrows. Now we need to take this to the next level. We need to get practice drawing the resonance structures when they are not given. To ease into it, we will still show the arrows, and we will focus on drawing the resonance structures with proper formal charges. Consider the following example:

In this example, we can see that one of the lone pairs on the oxygen is coming down to form a bond, and the C=C double bond is being pushed to form a lone pair on a carbon atom. When both arrows are pushed at the same time, we are not violating either of the two commandments. So, let's focus on how to draw the resonance structure. Since we know what arrows mean, it is easy to follow the arrows. We just get rid of one lone pair on the oxygen, place a double bond between the carbon and oxygen, get rid of the carbon–carbon double bond, and place a lone pair on the carbon:

The arrows are really a language, and they tell us what to do. But here comes the tricky part: we cannot forget to put formal charges on the new drawing. If we apply the rules of assigning formal charges, we see that the oxygen gets a positive charge and the carbon gets a negative charge. As long as we draw these charges, it is not necessary to draw in the lone pairs:

It is absolutely critical to draw these formal charges. Structures drawn without them are *wrong*. In fact, if you forget to draw the formal charges, then you are missing the whole point of resonance. Let's see why. Look at the resonance structure we just drew. Notice that there is a negative charge on a carbon atom. This tells us that this carbon atom is a site of high electron density. We would not know this by looking only at the first drawing of the molecule:

This is why we need resonance—it shows us where there are regions of high and low electron density. If we draw resonance structures without formal charges, then what is the point in drawing the resonance structures at all?

Now that we see that proper formal charges are essential, we should make sure that we know how to draw them when drawing resonance structures. If you are a little bit shaky when it comes to formal charges, go back and review formal charges in the previous chapter. But we can also see where to put formal charges without having to count each time. We saw the common situations for oxygen, nitrogen, and carbon. It is important to remember those (go back and review those if you need to).

Another way to assign formal charges is to read the arrows properly. Let's look at our example again:

Notice what the arrows are telling us: oxygen is giving up a lone pair (two electrons entirely on the oxygen) to form a bond (two electrons being shared: one for the oxygen and one for the carbon). So oxygen is losing an electron. This tells us that it must get a positive charge in the resonance structure. A similar analysis for the carbon on the bottom right shows that it will get a negative charge. Remember that the electrons are not really moving anywhere. Arrows are just tools that help us draw resonance structures. To use these tools properly, we imagine that the electrons are moving, but they are not.

Now let's practice.

EXERCISE 2.20 Draw the resonance structure that you get when you push the arrows shown below. Be sure to include formal charges.

Answer We read the arrows to see what is happening. One of the lone pairs on the oxygen is coming down to form a bond, and the C=C double bond is being pushed to form a lone pair on a carbon atom. This is very similar to the example we just saw.

We just get rid of one lone pair on the oxygen, place a double bond between the carbon and oxygen, get rid of the carbon–carbon double bond, and place a lone pair on the carbon. Finally, we must put in any formal charges:

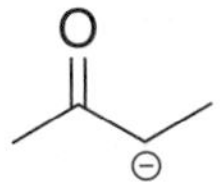

There is one subtle point to mention. We said that you do not need to draw lone pairs—you only need to draw formal charges. There will be times when you will see arrows being pushed on structures that do not have the lone pairs drawn. When this happens, you might see an arrow coming from a negative charge:

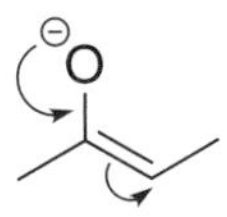

is the same as

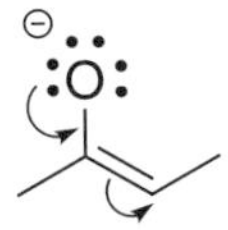

The drawing on the left is the common way this is drawn. Just don't forget that the electrons are really coming from a lone pair (as seen in the drawing on the right).

One way to double check your drawing when you are done is to count the total charge on the resonance structure that you draw. This total charge should be the same as the the structure you started with. So if the first structure has a negative charge, then the resonance structure you draw should also have a negative charge. If it doesn't, then you know you did something wrong (this is known as *conservation of charge*). You cannot change the total charge on a compound when drawing resonance structures.

PROBLEMS   For each of the structures below, draw the resonance structure that you get when you push the arrows shown. Be sure to include formal charges. (*Hint:* In some cases the lone pairs are drawn and in other cases they are not drawn. Be sure to take them into account even if they are not drawn—you need to train yourself to see lone pairs when they are not drawn.)

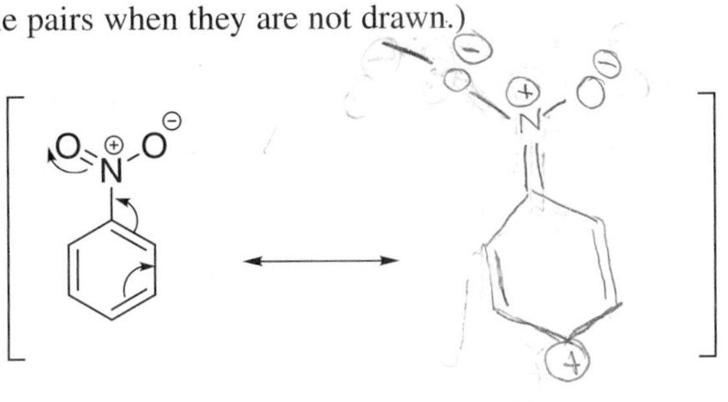

2.21

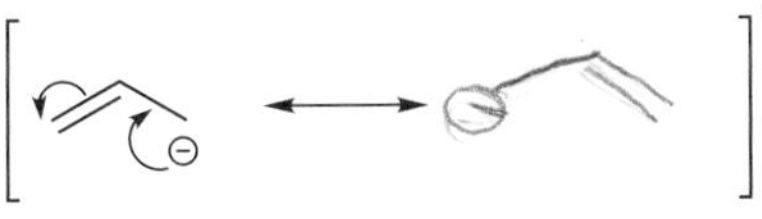

2.22

2.23

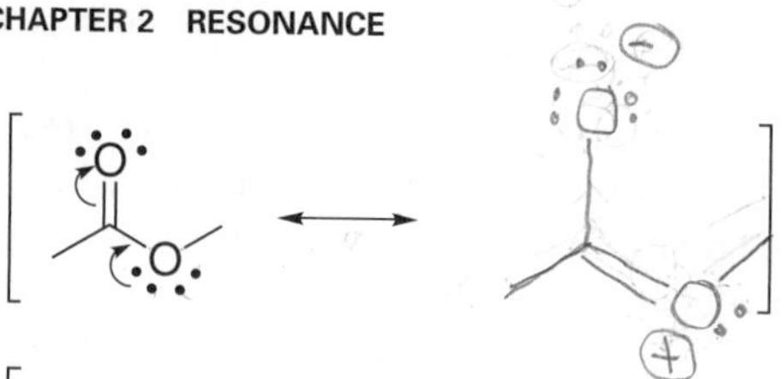

2.24

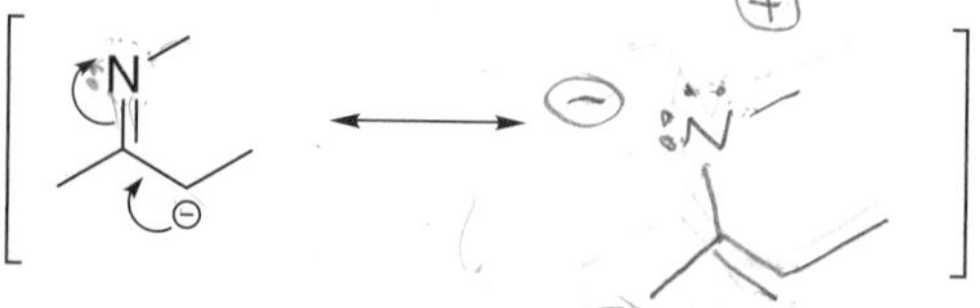

2.25

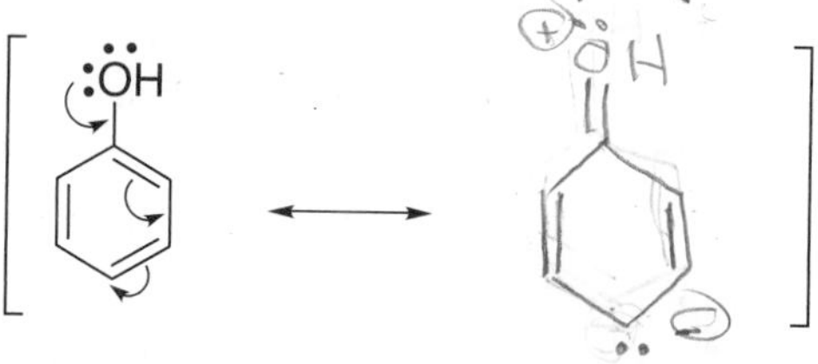

2.26

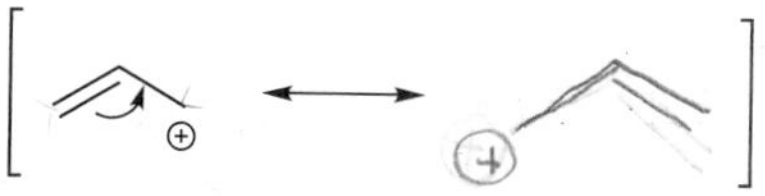

2.27

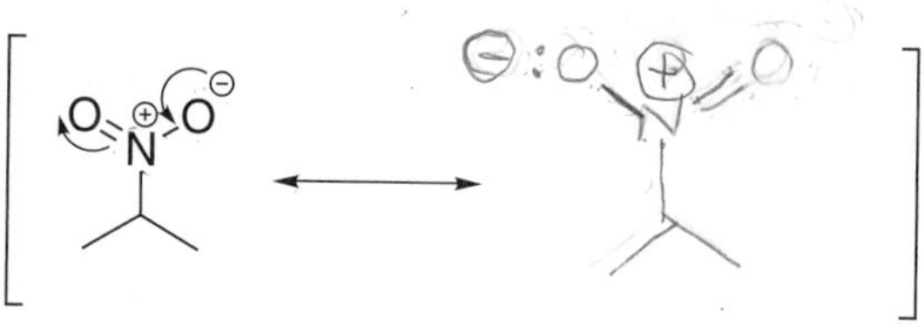

2.28

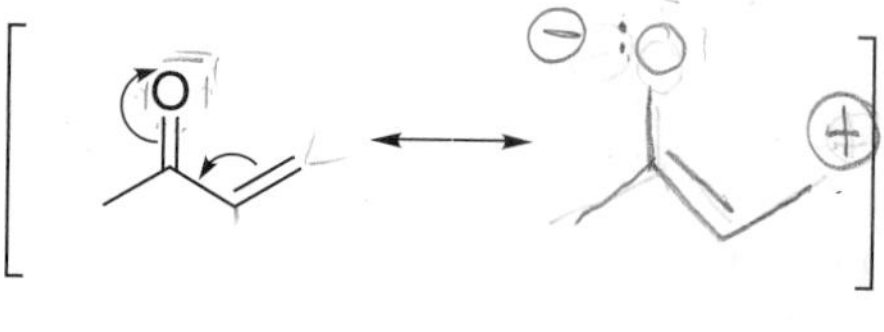

2.6 DRAWING RESONANCE STRUCTURES—STEP BY STEP

Now we have all the tools we need. We know why we need resonance structures and what they represent. We know about what curved arrows are and where not to draw them. We know how to recognize bad arrows that violate the two commandments. We know how to draw arrows that get you from one structure to another, and we know how to draw in formal charges. We are now ready for the final challenge: drawing curved arrows when we do not know what the next resonance structure looks like. Now that you know when you can and cannot push arrows, you need to practice using arrow pushing to determine by yourself how to draw the other resonance structures.

First we need to locate the part of the molecule where resonance is an issue. Remember that we can push electrons only from lone pairs or bonds. We don't need to worry about all bonds, because we can't push an arrow from a single bond (that would violate the first commandment). So we only care about double or triple bonds. Double and triple bonds are called *pi bonds*. So we need to look for lone pairs and pi bonds. Usually, only a small region of the molecule will possess either of these features.

Once we have located the regions where resonance is an issue, now we need to ask if there is any way to push the electrons without violating the two commandments. Let's be methodical, and break this up into three questions:

1. Can we convert any *lone pairs into pi bonds* without violating the two commandments?

2. Can we convert any *pi bonds into lone pairs* without violating the two commandments?

3. Can we convert any *pi bonds into pi bonds* without violating the two commandments?

We do not need to worry about the fourth possibility (converting a lone pair into a lone pair) because electrons cannot jump from one atom to another. Only the three possibilities above are acceptable.

Let's go through these three steps, one at a time, starting with step 1, converting lone pairs into bonds. Consider the following example:

We ask if there are any lone pairs that we can move to form a pi bond. So we draw an arrow that brings the lone pair down to form a pi bond:

This does not violate either of the two commandments. We did not break any single bonds and we did not violate the octet rule. So this is a valid structure. Notice that we cannot move the lone pair in another direction, because then we would be violating the octet rule:

Let's try again with the following example:

We ask if we can move one of the lone pairs down to form a pi bond, so we try to draw it:

This violates the octet rule—the carbon atom would end up with five bonds. So we cannot push the arrows that way. There is no way to turn the lone pair into a pi bond in this example.

Now let's move on to step 2, converting pi bonds into lone pairs. We try to move the double bond to form a lone pair and we see that we can move the bond in either direction:

or

Neither of these structures violates the two commandments, so both structures above are valid resonance structures. (However, the bottom structure, although valid, is not a significant resonance structure. In the next section, we will see how to determine which resonance structures are significant and which are not.)

For step 3, converting pi bonds into pi bonds, let's consider the following examples:

If we try to push the pi bonds to form other pi bonds, we find

No: This violates the octet rule.

Yes: Does not violate the octet rule.

The top structure violates the octet rule (giving carbon five bonds), and the bottom structure does not violate the octet rule. The arrow on the bottom structure will therefore provide a valid resonance structure:

Now that we have learned all three steps, we need to consider that these steps can be combined. Sometimes we cannot do a step without violating the octet rule, but by doing two steps at the same time, we can avoid violating the octet rule. For example, if we try to turn a lone pair into a bond in the following structure, we see that this would violate the octet rule:

If, at the same time, we also do step 2 (push a pi bond to become a lone pair), then it works:

In other words, you should not always jump to the conclusion that pushing an arrow will violate the octet rule. You should first look to see if you can push another arrow that will eliminate the problem.

As another example, consider the structure below. We cannot move the $C{=}C$ bond to become another bond unless we also move the $C{=}O$ bond to become a lone pair:

No   Yes

In this way, we truly are "pushing" the electrons around.

Now we are ready to get some practice drawing resonance structures.

EXERCISE 2.29   Draw all resonance structures for the following compound:

Answer Let's start by finding all of the lone pairs and redrawing the molecule. Oxygen has two bonds here, so it must have two lone pairs (so that it will be using all four orbitals):

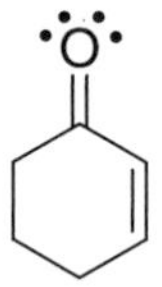

Now let's do step 1: can we convert any lone pairs into pi bonds? If we try to bring down the lone pairs, we will violate the octet rule by forming a carbon atom with five bonds:

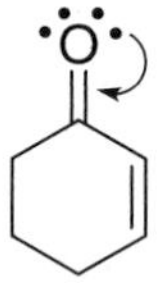

Violates second commandment

The only way to avoid forming a fifth bond for carbon would be to push an arrow that takes electrons away from that carbon. If we try to do this, we will break a single bond and we will be violating the first commandment:

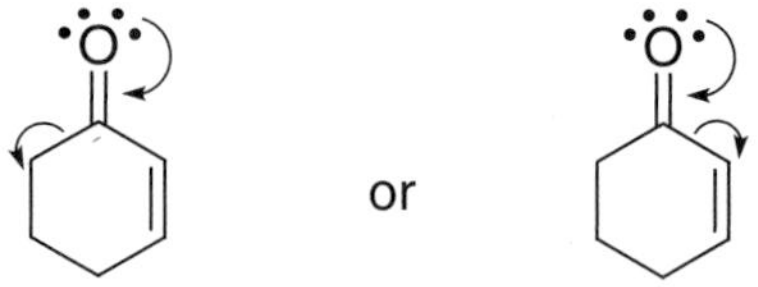

or

Violates first commandment

We cannot move a lone pair to form a pi bond, so we move on to step 2: can we convert any pi bonds into lone pairs? Yes:

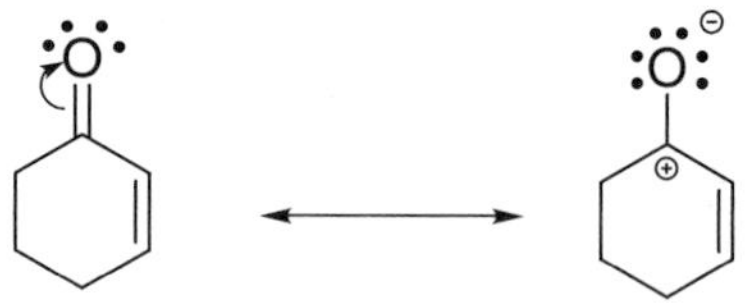

Now we move to step 3: can we convert pi bonds into pi bonds? There is only one move that will not violate the two commandments:

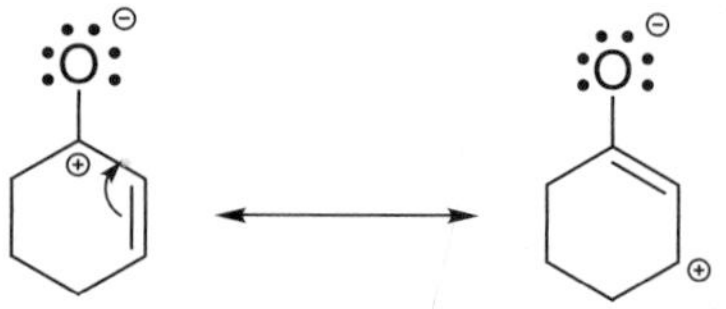

So the resonance structures are

PROBLEM 2.30 For the following compound, go through all three steps (making sure not to violate the two commandments) and draw the resonance structures.

While working through this problem, you probably found that it took a very long time to think through every possibility, to count lone pairs, to worry about violating the octet rule for each atom, to assign formal charges, and so on. Fortunately, there is a way to avoid all of this tedious work. You can learn how to become very quick and efficient at drawing resonance structures if you learn certain patterns and train yourself to recognize those patterns. We will now develop this skill.

2.7 DRAWING RESONANCE STRUCTURES— BY RECOGNIZING PATTERNS

There are five patterns that you should learn to recognize to become proficient at drawing resonance structures. First we list them, and then we will go through each pattern in detail, with examples and exercises. Here they are:

1. A lone pair next to a pi bond.
2. A lone pair next to a positive charge.
3. A pi bond next to a positive charge.
4. A pi bond between two atoms, where one of those atoms is electronegative.
5. Pi bonds going all the way around a ring.

A Lone Pair Next to a Pi Bond

Let's see an example before going into the details:

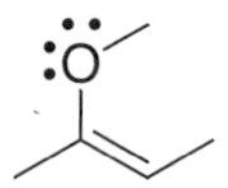

The atom with the lone pair can have no formal charge (as above), or it can have a negative formal charge:

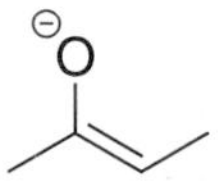

The important part is having a lone pair "next to" the pi bond. "Next to" means that the lone pair is separated from the double bond by exactly one single bond—no more and no less. You can see this in all of the examples below:

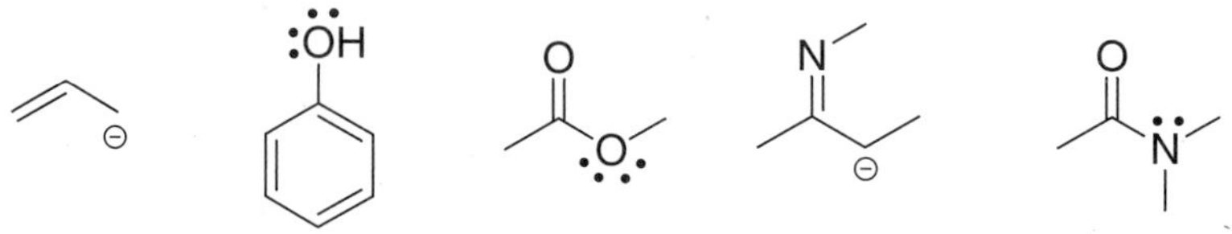

In each of these cases, you can bring down the lone pair to form a pi bond, and kick up the pi bond to form a lone pair:

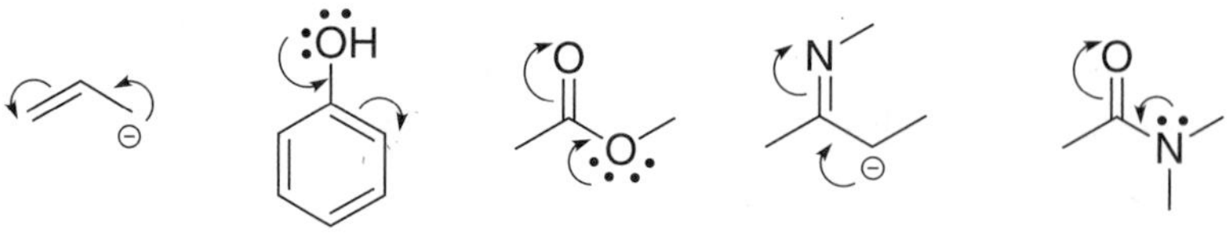

Notice what happens with the formal charges. When the atom with the lone pair has a negative charge, then it transfers its negative charge to the atom that will get a lone pair in the end:

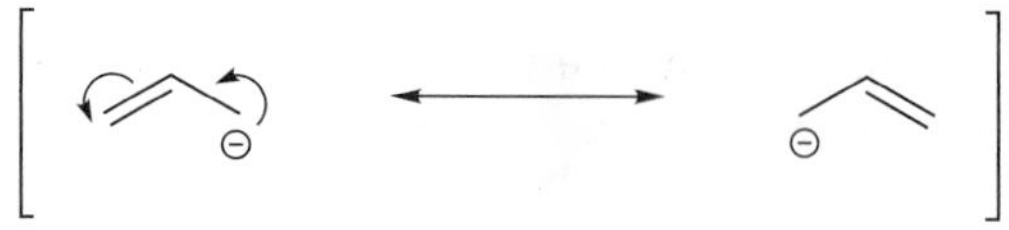

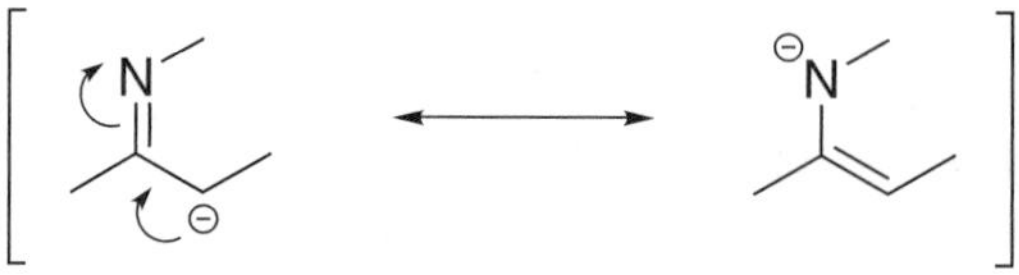

When the atom with the lone pair does not have a negative charge to begin with, then it will end up with a positive charge in the end, while a negative charge will go on the atom getting the lone pair in the end (remember conservation of charge):

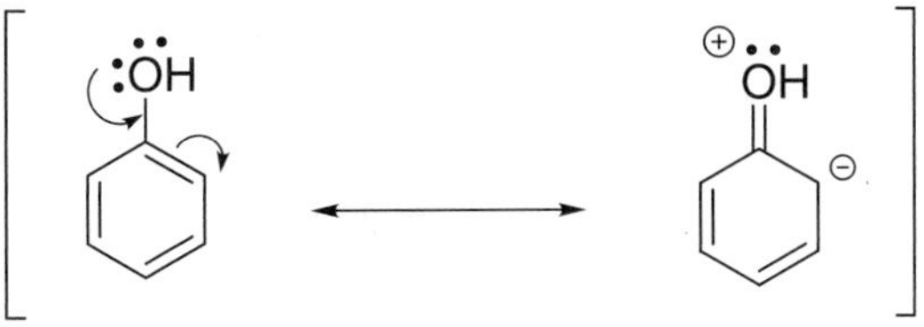

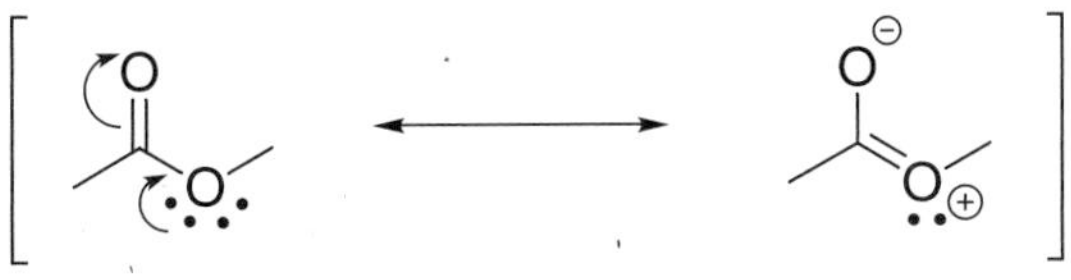

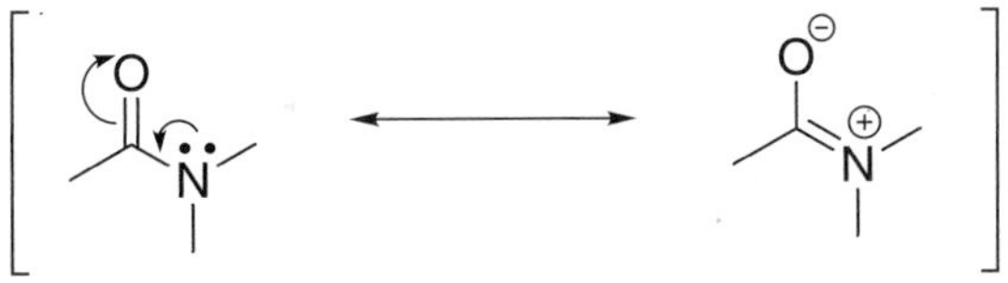

Once you learn to recognize this pattern (a lone pair next to a pi bond), you will be able to save time in calculating formal charges and determining if the octet rule is being violated. You will be able to push the arrows and draw the new resonance structure without thinking about it.

EXERCISE 2.31   Draw the resonance structure of the compound below:

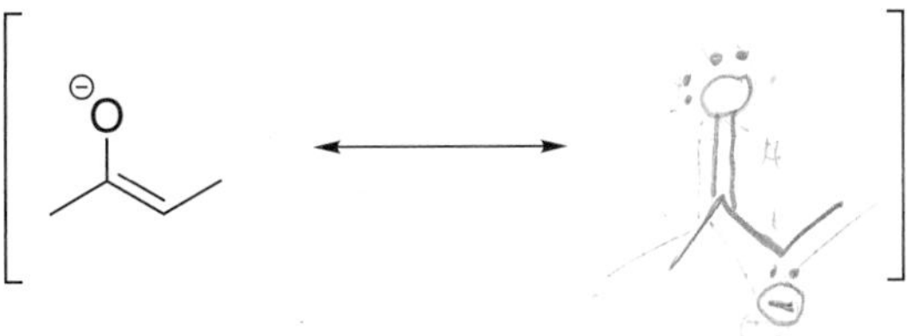

Answer   We notice that this is a lone pair next to a pi bond. Therefore, we push two arrows: one from the lone pair to form a pi bond, and one from the pi bond to form a lone pair.

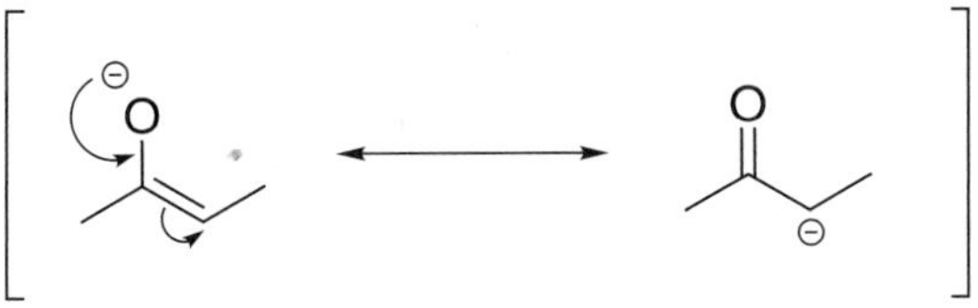

Look carefully at the formal charges. The negative charge used to be on the oxygen, but now it moved to the carbon.

PROBLEMS For each of the compounds below, locate the pattern we just learned and draw the resonance structure.

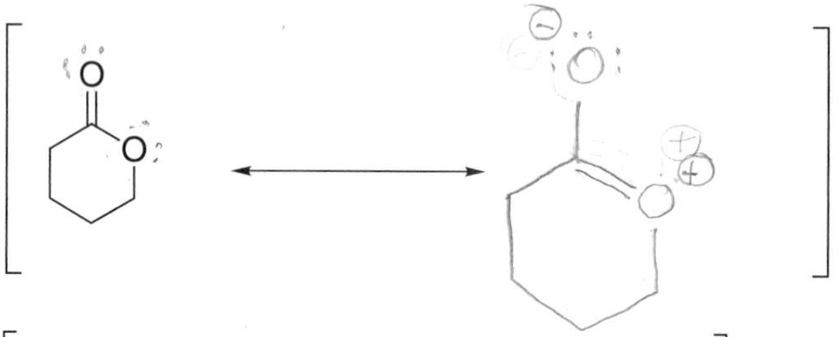

2.32

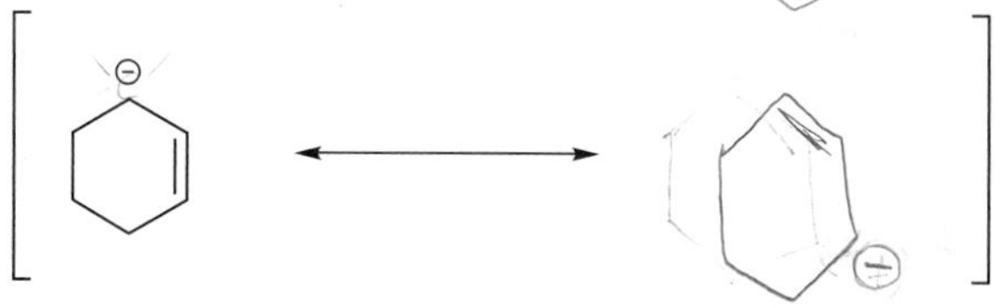

2.33

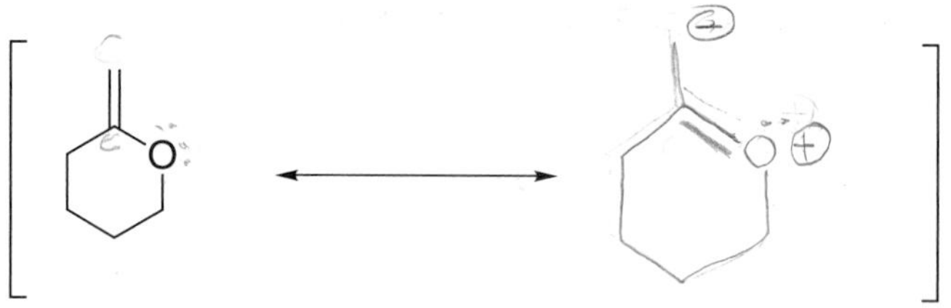

2.34

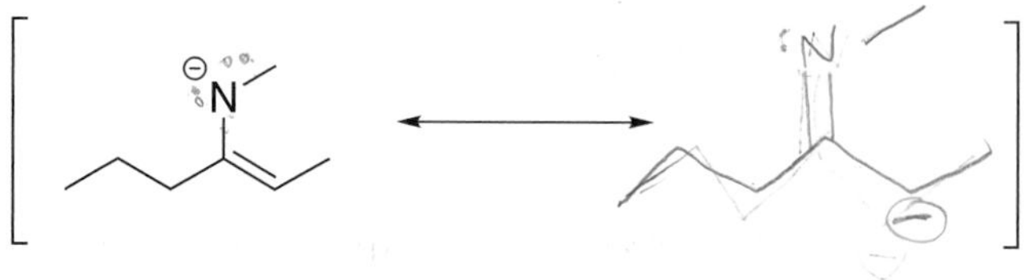

2.35

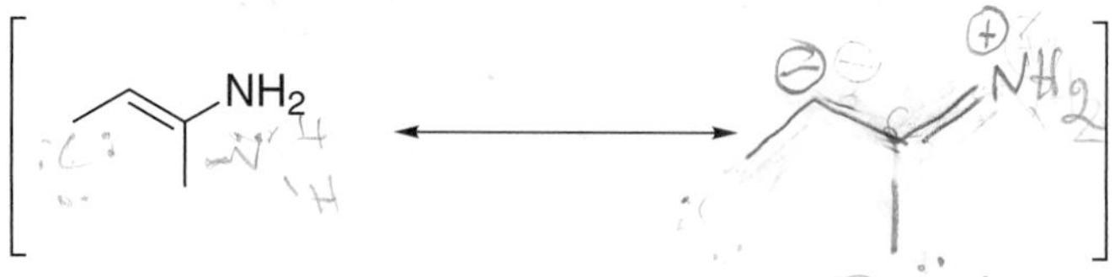

2.36

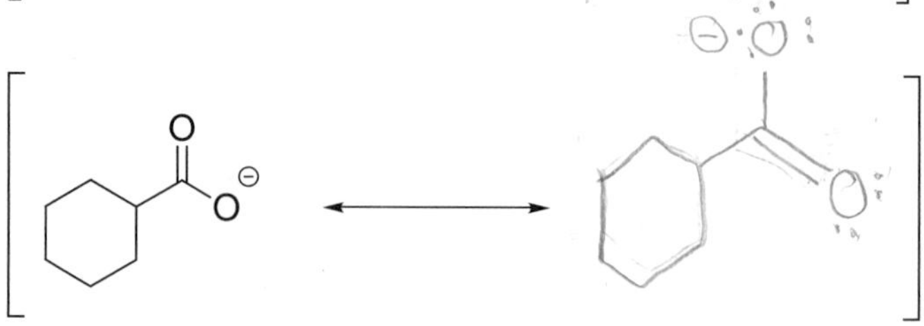

2.37

2.38

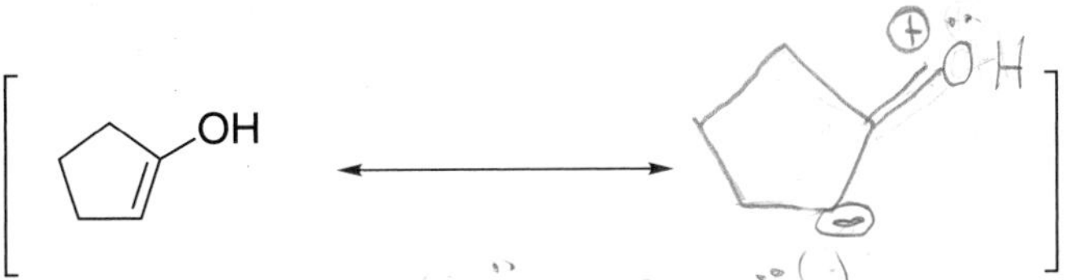

2.39

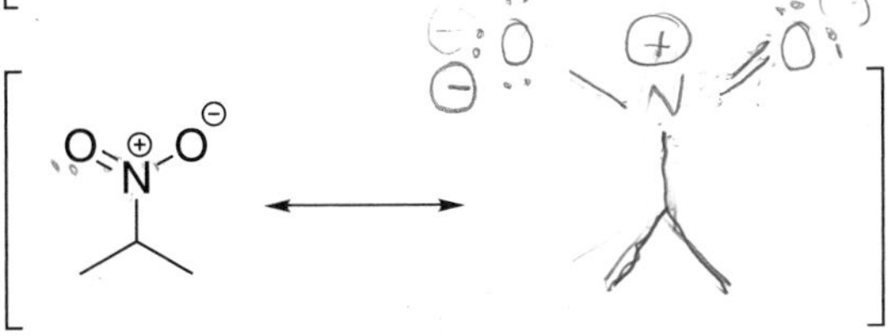

Notice that the lone pair needs to be directly next to the pi bond. If we move the lone pair one atom away, this does not work anymore:

Good   Bad

A Lone Pair Next to a Positive Charge

Let's see an example:

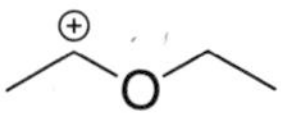

The atom with the lone pair can have no formal charge (as above) or it can have a negative formal charge:

The important part is having a lone pair next to a positive charge. In each of the above cases, we can bring down the lone pair to form a pi bond:

Notice what happens with the formal charges. When the atom with the lone pair has a negative charge, then the charges end up canceling each other:

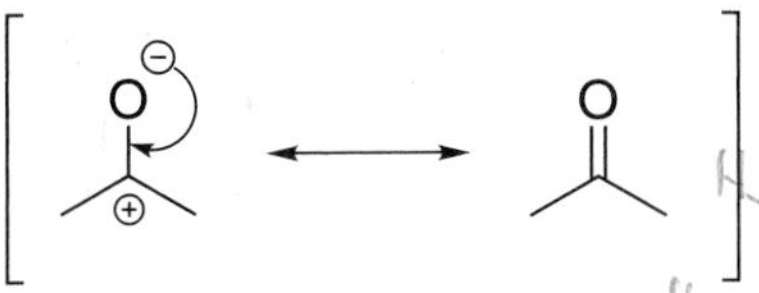

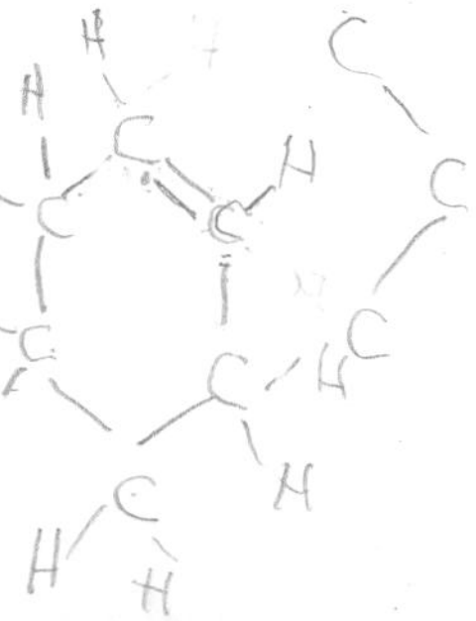

When the atom with the lone pair does not have a negative charge to begin with, then it will end up with the positive charge in the end (remember conservation of charge):

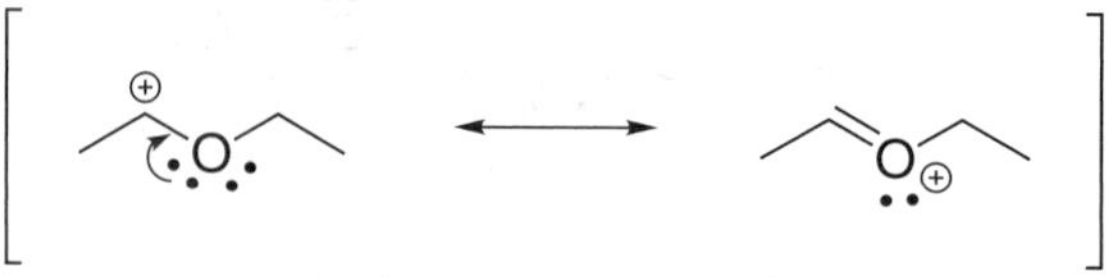

PROBLEMS For each of the compounds below, locate the pattern we just learned and draw the resonance structure.

2.40

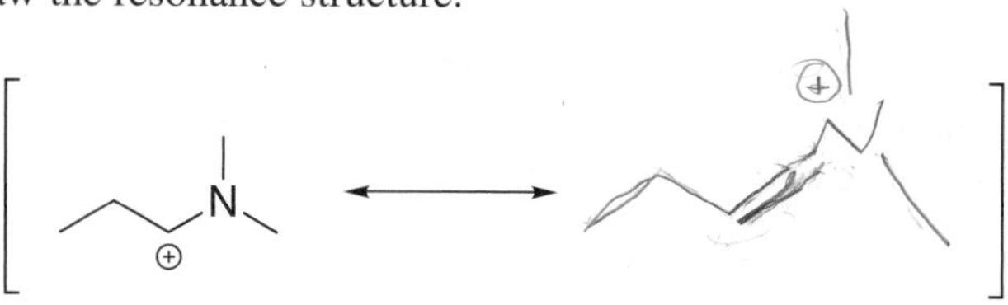

2.41

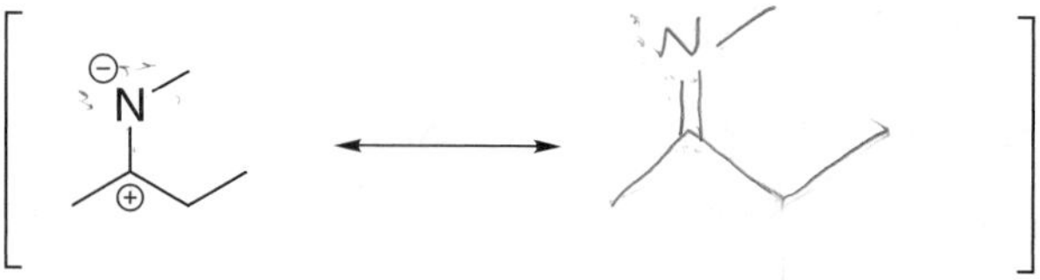

2.42

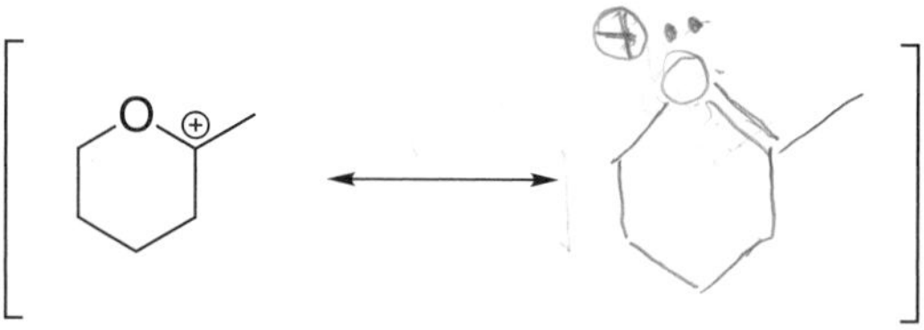

2.43

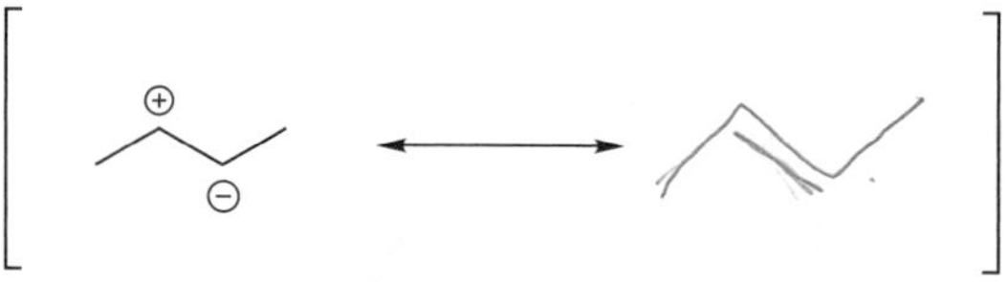

Notice that in this problem, a negative and positive charge cancel each other to become a double bond. There is one situation when we cannot combine charges to give a double bond: the nitro group. The structure of the nitro group looks like this:

We cannot draw a resonance structure where there are no charges:

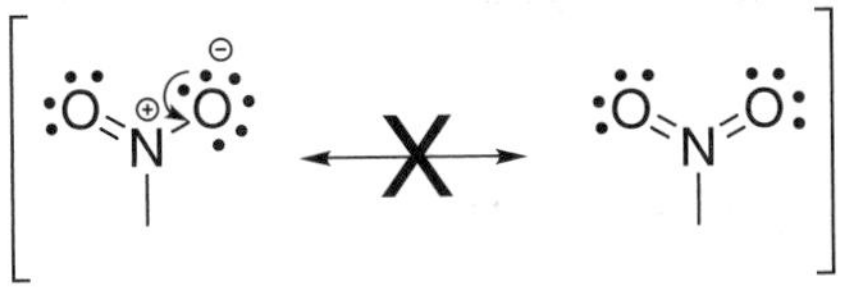

This might seem better at first, because we get rid of the charges, but our two commandments show us why it cannot be drawn like this: the nitrogen atom would have five bonds, which this would violate the octet rule.

A Pi Bond Next to a Positive Charge

These cases are very easy to see:

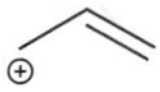

We need only one arrow going from the pi bond to form a new pi bond:

Notice what happens to the formal charge in the process. It gets moved to the other end:

It is possible to have many double bonds in conjugation (this means that we have many double bonds that are each separated by only one single bond) next to a positive charge:

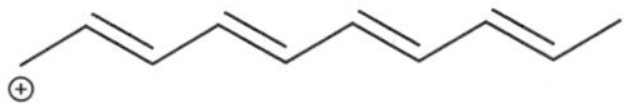

When this happens, we can push all of the double bonds over, and we don't need to worry about calculating formal charges—just move the positive charge to the other end:

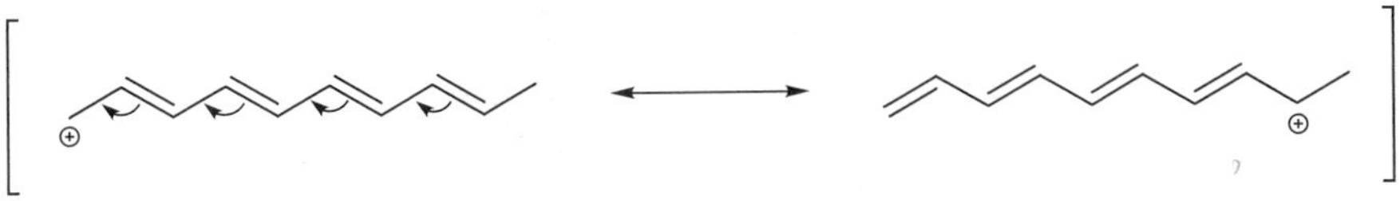

Of course, we should push one arrow at a time so that we can draw *all* of the resonance structures. But it is nice to know how the formal charges will end up so that we don't have to calculate them every time we push an arrow.

PROBLEMS   For each of the compounds below, locate the pattern we just learned
and draw the resonance structure.

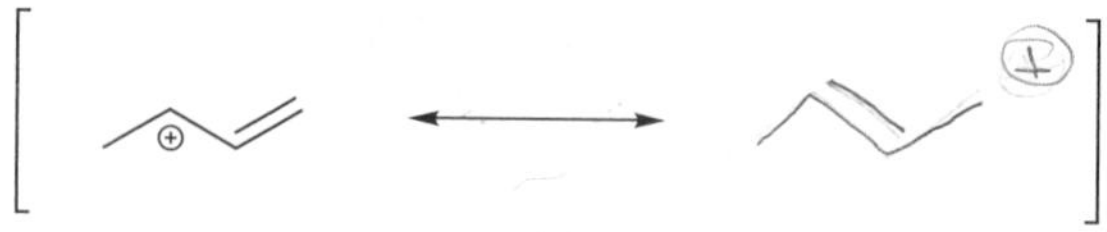

2.44

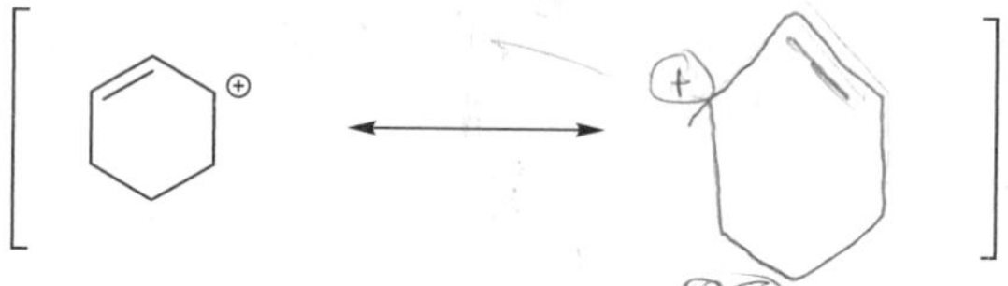

2.45

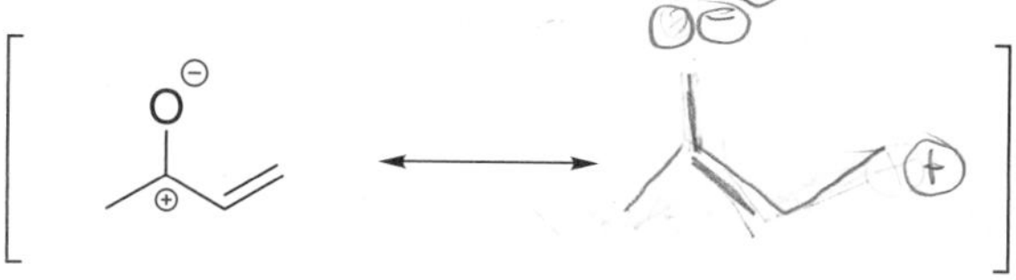

2.46

A Pi Bond Between Two Atoms, Where One of Those Atoms Is Electronegative (N, O, etc.)

Let's see an example:

In cases like this, we move the pi bond up onto the electronegative atom to become
a lone pair:

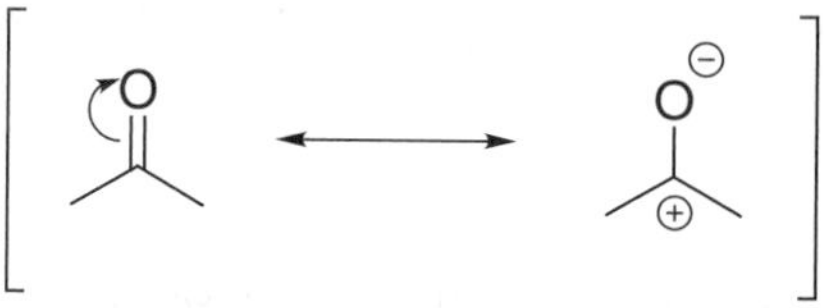

Notice what happens with the formal charges. A double bond is being separated into
a positive and negative charge (this is the opposite of what we saw in the second pattern we looked at, where the charges came together to form a double bond).

PROBLEMS   For each of the compounds below, locate the pattern we just learned
and draw the resonance structure:

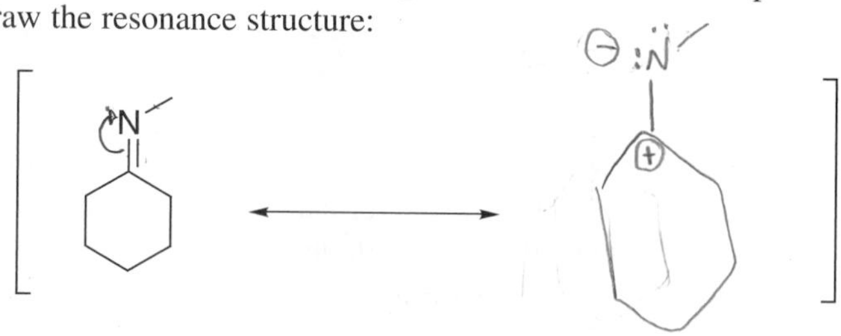

2.47

2.48

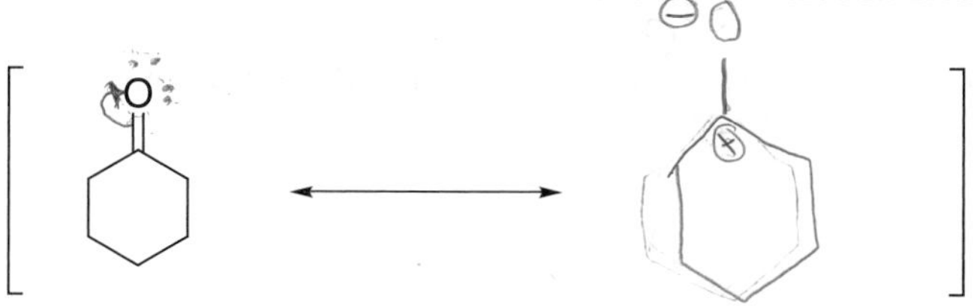

2.49

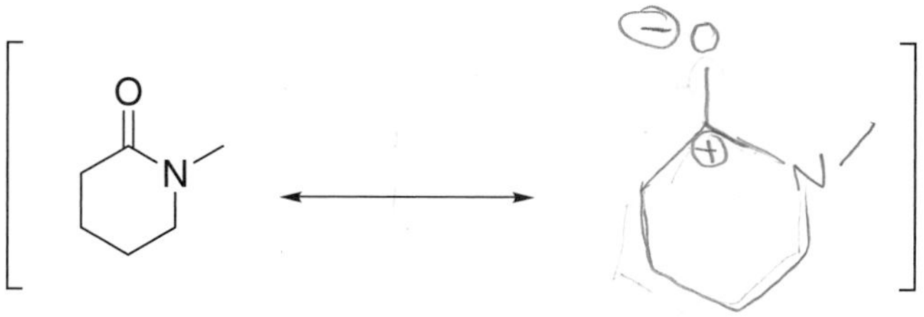

Pi Bonds Going All the Way Around a Ring

Whenever we have alternating double and single bonds, we refer to the alternating bond system as *conjugated:*

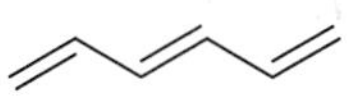

Conjugated double bonds

When we have a conjugated system that wraps around in a circle, then we can always move the electrons around in a circle:

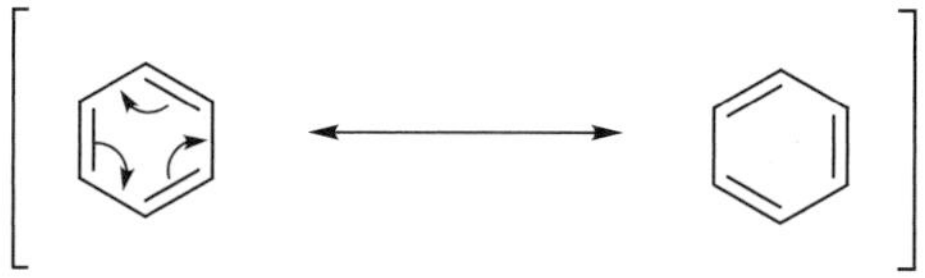

It does not matter whether we push our arrows clockwise or counterclockwise (either way gives us the same result, and remember that the electrons are not really moving anyway).

Now we are ready to go back to some problems. Let's try to draw resonance structures again. Only this time, let's try to focus on recognizing some patterns. Look at the examples below, and see if you can recognize any of the patterns we just discussed:

1. A lone pair next to a pi bond.
2. A lone pair next to a positive charge.
3. A pi bond next to a positive charge.
4. A pi bond between two atoms, where one of those atoms is electronegative.
5. Pi bonds going all the way around a ring.

PROBLEMS For each of the following compounds, draw the resonance structures.

2.50

2.51

2.52

2.53

2.54

2.55

2.56

2.57

2.58

2.59

2.60

2.8 ASSESSING THE RELATIVE IMPORTANCE OF RESONANCE STRUCTURES

Not all resonance structures are always important. Although there are often many valid resonance structures for a single compound (meaning that those structures do not violate the two commandments), usually many of these resonance structures are not significant. For example, the following resonance structures are not significant even though they are valid structures:

There are three simple rules to follow to determine which resonance structures are significant. At this point, you are probably thinking that it is hard enough to keep track of everything we have seen so far—there are two commandments for how not to push arrows, there are three steps for determining valid resonance structures, and now there are three rules for determining which resonance structures are significant. The good news is that this is the end of the line. There are no more rules or steps. We are almost done with resonance structures. More good news—drawing resonance structures really is very much like riding a bike. When you first learn to ride a bike, you need to concentrate on every movement to avoid from falling. And you have to remember a lot of rules, such as which way to lean your body and which way to turn the steering wheel when you feel you are falling to the left. But eventually, you get the hang of it, and then you can do it with no hands. The same is true here. It will take a lot of practice. But before you know it, you will be the resonance guru, and that is where you need to be to do well in this class.

Let's see the three rules for finding which resonance structures are significant.

Rule 1   Minimize charges. The best kind of structure is one without any charges. It is OK to have two charges, but you should try to avoid structures that have more than two charges. Compare the following two cases:

Both compounds have a pi bond between a carbon atom and an electronegative atom (C=O), and both compounds have a lone pair next to the pi bond. So we would expect their resonance structures to be similar, and we would expect these compounds to have the same number of resonance structures. But they do not. Let's see why. Let's start by drawing the resonance structures of the first compound:

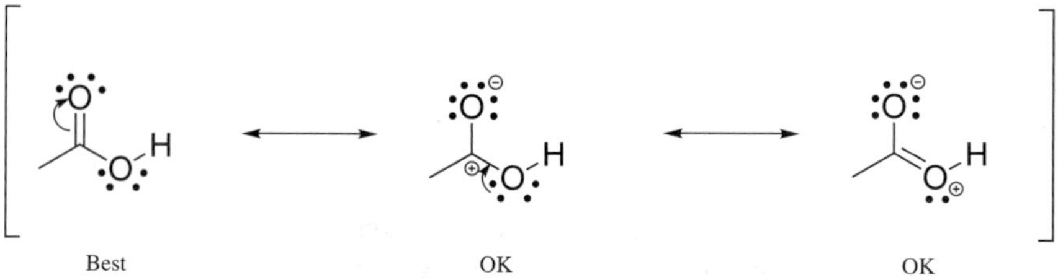

The first resonance structure is the best, because there is no charge separation. The other two drawings have charge separation, but there are only two charges in each drawing and the negative charge is on an oxygen. (Rule 2 explains why it is important for the charge to be on oxygen.)

Now, let's do the same thing for the other compound:

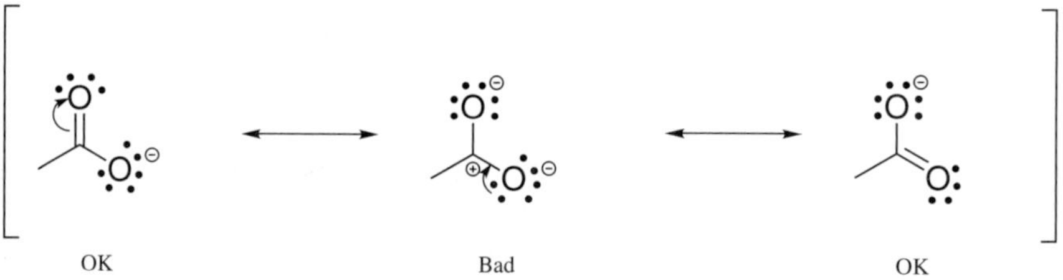

The first and last structures are OK, but the second resonance structure is bad because there are too many charges. So we don't draw it when drawing resonance structures. It is not significant.

Rule 2 It is OK to convert a structure with no formal charge into a structure with a negative charge and a positive charge as long as the negative charge is on an electronegative atom. For example, it is OK to draw the following:

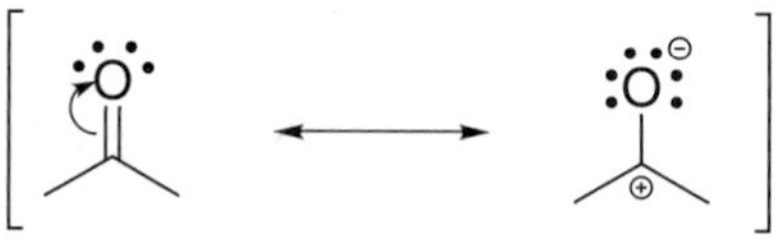

But, we should not draw it this way:

Technically, this is a valid resonance structure, since we have not violated the two commandments. Nevertheless, we do not consider it to be significant because the negative charge is on carbon and the positive charge is on oxygen.

Similarly, we would not draw the following resonance structure for a double bond:

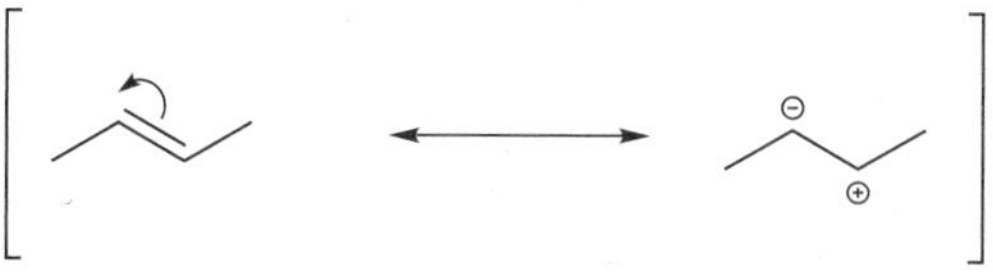

This is a valid structure, but we have created a charge separation with no apparent benefit (the negative charge is not on an electronegative atom), so it is not significant.

Rule 3   There are cases when we can draw a resonance structure with a positive charge on an electronegative atom. We can do this if it gives all of the atoms an octet of electrons. For example, consider the following structures:

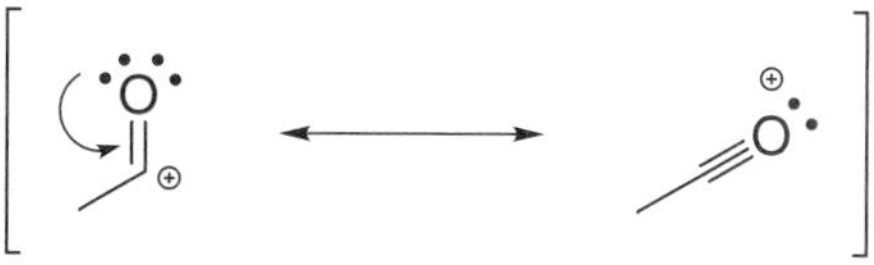

In the first structure, the oxygen has its octet, but the carbon only has six electrons. When we draw the resonance structure, oxygen has a positive charge, but both oxygen and carbon have their octet. So this is a good resonance structure even though oxygen has a positive charge.

Here is another example with a positive charge on nitrogen:

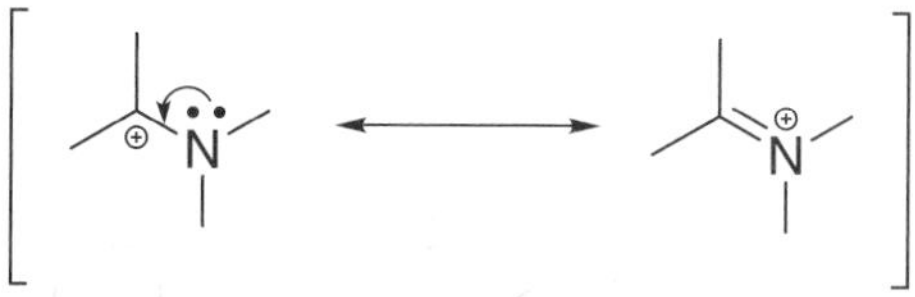

In the first structure, the nitrogen has its octet, but the carbon only has six electrons. When we draw the resonance structure, nitrogen has a positive charge, but notice that both nitrogen and carbon have their octet. So this is a good resonance structure even though nitrogen has a positive charge.

We can put a positive charge on an electronegative atom only if doing so will give all atoms an octet. We cannot do it in the following case:

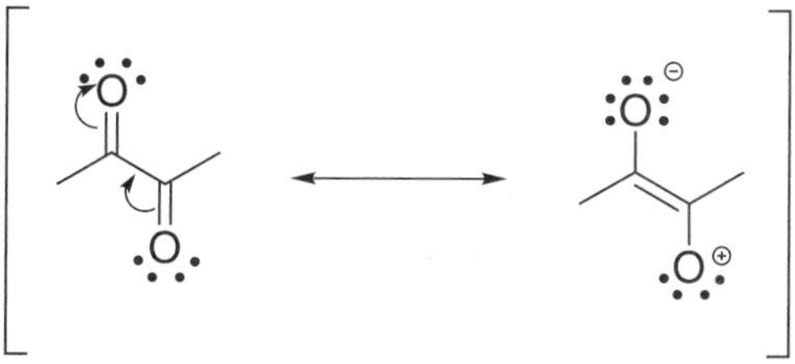

The oxygen with the positive charge does not have its octet, so this resonance structure is not significant.

PROBLEMS For each of the following compounds, draw all of the *significant* resonance structures.

2.61

2.62

2.63

2.64

2.65

2.66

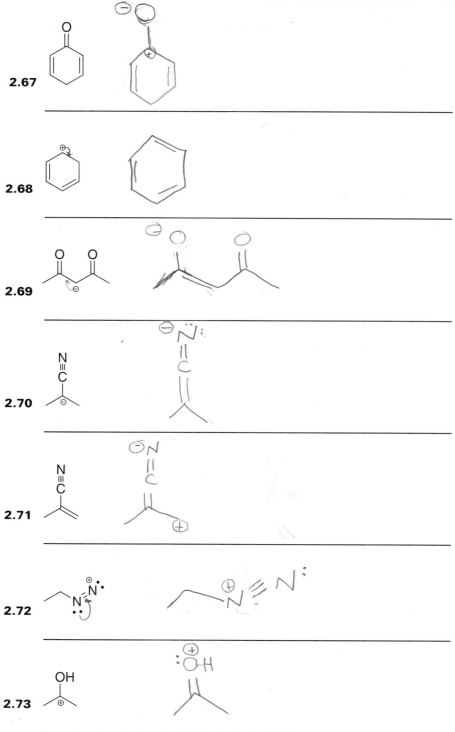

2.67

2.68

2.69

2.70

2.71

2.72

2.73

2.74

2.75

ACID–BASE REACTIONS

The first several chapters of any organic chemistry textbook focus on the structure of molecules: how atoms connect to form bonds, how we draw those connections, the problems with our drawing methods, how we name molecules, what molecules look like in 3D, how molecules twist and bend in space, and so on. Only after gaining a clear understanding of structure do we move on to reactions. But there seems to be one exception: acid–base chemistry.

Acid–base chemistry is typically covered in one of the first few chapters of organic chemistry textbook, yet it might seem to belong better in the later chapters on reactions. There is an important reason why acid–base chemistry is taught so early on in your course. By understanding this reason, you will have a better perspective of why acid–base chemistry is so incredibly important.

To appreciate the reason for teaching acid–base chemistry early in the course, we need to first have a very simple understanding of what acid–base chemistry is all about. Let's summarize with a simple equation:

$$HA \rightleftharpoons H^+ + A^-$$

In the equation above, we see an acid (HA) on the left side of the equilibrium, and the conjugate base (A^-) on the right side. HA is an acid by virtue of the fact that it has a proton (H^+) to give. A^- is a base by virtue of the fact that it wants to take its proton back (acids give protons and bases take protons). Since A^- is the base that we get when we deprotonate HA, we call A^- the *conjugate base* of HA.

So the question is: how much is HA willing to give up its proton? If HA is very willing to give up the proton, then HA is a strong acid. However, if HA is not willing to give up its proton, then HA is a weak acid. So, how can we tell whether or not HA is willing to give up its proton? *We can figure it out by looking at the conjugate base.*

Notice that the conjugate base has a negative charge. The real question is: how stable is that negative charge? If that charge is stable, then HA will be willing to give up the proton, and therefore HA will be a strong acid. If that charge is not stable, then HA will not be willing to give up its proton, and HA will be a weak acid.

So you only need one skill to completely master acid–base chemistry: you need to be able to look at a negative charge and determine how stable that negative charge is. If you can do that, then acid–base chemistry will be a breeze for you. If you cannot determine charge stability, then you will have problems even after you finish acid–base chemistry. To predict reactions, you need to know what kind of charges are stable and what kind of charges are not stable.

Now you can understand why acid–base chemistry is taught so early in the course. Charge stability is a vital part of understanding the structure of molecules. It is so incredibly important because reactions are all about how charges interact with one another. You cannot begin to discuss reactions until you have an excellent understanding of what factors stabilize charges and what factors destabilize charges. This chapter will focus on the four most important factors, one at a time.

3.1 FACTOR 1—WHAT ATOM IS THE CHARGE ON?

The most important factor for determining charge stability is to ask what atom the charge is on. For example, consider the two charged compounds below:

The one on the left has a negative charge on oxygen, and the one on the right has the charge on sulfur. How do we compare these? We look at the periodic table, and we need to consider two trends: comparing atoms *in the same row* and comparing atoms *in the same column:*

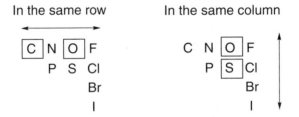

Let's start with comparing atoms *in the same row.* For example, let's compare carbon and oxygen:

The compound on the left has the charge on carbon, and the compound on the right has the charge on oxygen. Which one is more stable? Recall that electronegativity increases as we move to the right on the periodic table:

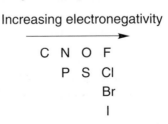

Since electronegativity is the measure of an element's affinity for electrons (how willing the atom will be to accept a new electron), we can say that a negative charge on oxygen will be more stable than a negative charge on carbon.

Now let's compare atoms *in the same column,* for example, iodide (I⁻) and fluoride (F⁻). Here is where it gets a little bit tricky, because the trend is the opposite of the electronegativity trend:

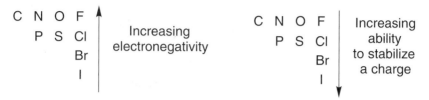

It is true that fluorine is more electronegative than iodine, but there is another more important trend when comparing atoms in the same column: the *size* of the atom. Iodine is *huge* compared to fluorine. So when a charge is placed on iodine, the charge is spread out over a very large volume. When a charge is placed on fluorine, the charge is stuck in a very small volume of space:

Even though fluorine is more electronegative than iodine, nevertheless, iodine can better stabilize a negative charge. If I⁻ is more stable than F⁻, then HI must be a stronger acid than HF, because HI will be more willing to give up its proton than HF.

To summarize, there are two important trends: *electronegativity* (for comparing atoms in the same row) and *size* (for comparing atoms in the same column). The first factor (comparing atoms in the same row) is a much stronger effect. In other words, the difference in stability between C⁻ and F⁻ is much greater than the difference in stability between I⁻ and F⁻.

Now we have all of the information we need to solve the first problem presented in this chapter: Which charge below is more stable?

When comparing these two ions, we see an oxygen atom bearing the negative charge (on the left) and a sulfur atom bearing the negative charge (on the right). Oxygen and sulfur are in the same column of the periodic table, so size is the important trend to look at. Sulfur is larger than oxygen, so sulfur can better stabilize the negative charge.

EXERCISE 3.1 Compare the two protons in the following compound. Which one is more acidic?

Answer We begin by pulling off one proton and drawing the conjugate base that we get. Then, we do the same thing for the other proton:

Now we need to compare these conjugate bases and ask which one is more stable. In other words, which negative charge is more stable? We are comparing a negative charge on nitrogen with a negative charge on oxygen. So we are comparing two atoms in the same row of the periodic table, and the important trend is electronegativity. Oxygen can better stabilize the negative charge, because oxygen is more electronegative than nitrogen. The proton on the oxygen will be more willing to come off, so it is more acidic:

PROBLEMS

3.2 Compare the two protons clearly shown in the following compound. (There are more protons in the compound, but only two are shown.) Which of these two protons is more acidic? Remember to begin by drawing the two conjugate bases, and then compare them.

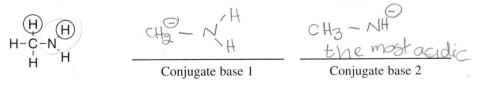

the most stable
acidic
Conjugate base 1   Conjugate base 2

3.3 Compare the two protons clearly shown in the following compound. Which of these two protons is more acidic?

the most acidic
Conjugate base 1   Conjugate base 2

3.4 Compare the two protons shown in the following compound, and identify which proton is more acidic:

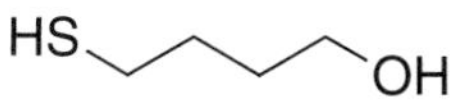

HS⌣⌣⌣OH   S⌣⌣⌣OH _the most acidic_   HS⌣⌣⌣O⁻

_____   _____
Conjugate base 1 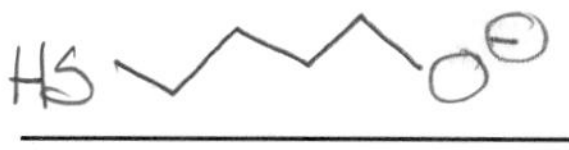   Conjugate base 2

3.5 Compare the two protons shown in the following compound, and identify which proton is more acidic:

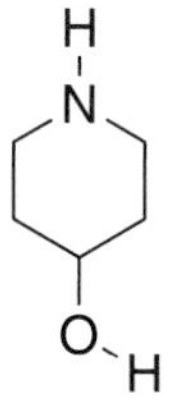

[structure: piperidine with N–H and O–H]   [structure: Conjugate base 1 with N⁻ and OH]   [structure: Conjugate base 2 with N–H and O⁻]

_____   _____
Conjugate base 1 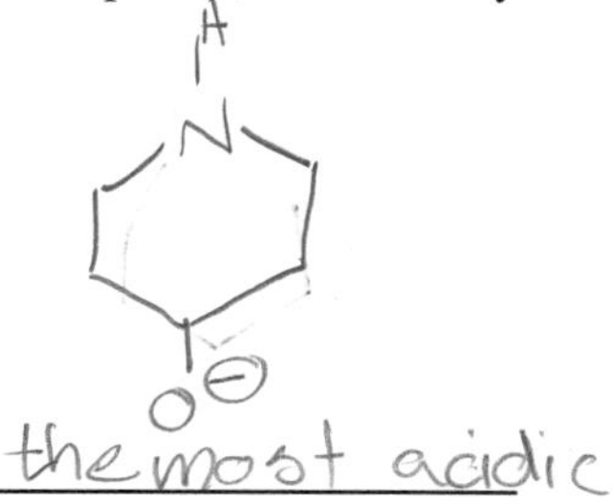   Conjugate base 2   _the most acidic_

3.2   FACTOR 2—RESONANCE

The last chapter was devoted solely to drawing resonance structures. If you have not yet completed that chapter, do so before you begin this section. We said in the last chapter that resonance would find its way into every single topic in organic chemistry. And here it is in acid–base chemistry.

To see how resonance plays a role here, let's compare the following two compounds:

[structures: acetic acid and ethanol/ether with O–H]

In both cases, if we pull off the proton, we get a charge on oxygen:

[structures: acetate anion and ethoxide anion]

So we cannot use factor 1 (what atom is the charge on) to determine which proton is more acidic. In both cases, we are dealing with a negative charge on oxygen. But there is a critical difference between these two negative charges. The one on the left is stabilized by resonance:

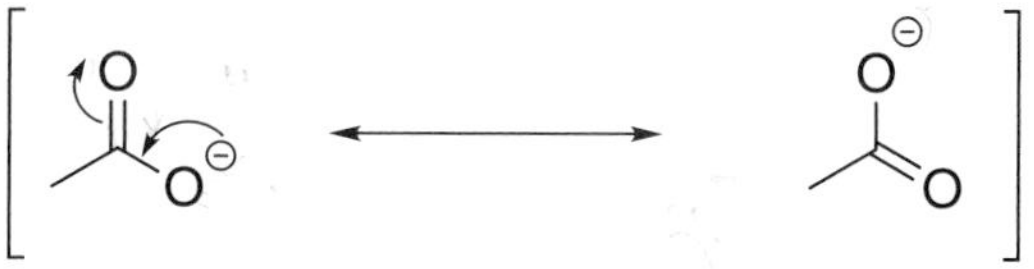

[resonance structures of acetate in brackets]

Remember what resonance means. It does not mean that we have two structures that are in equilibrium. Rather, it means that there is only one compound, and we cannot use one drawing to adequately describe where the charge is. In reality, the charge is spread out equally over both oxygen atoms. To see this we need to draw both drawings.

So what does this do in terms of stabilizing the negative charge? Imagine that you have a hot potato in your hand (too hot to hold for long). If you could grab another potato that is cold and transfer half of the warmth to the second potato, then you would have two potatoes, each of which is not too hot to hold. It's the same concept here. When we spread a charge over more than one atom, we call the charge "delocalized." A delocalized negative charge is more stable than a localized negative charge (stuck on one atom):

Charge is stuck on one atom ("localized")

This factor is very important and explains why carboxylic acids are acidic:

They are acidic because the conjugate base is stabilized by resonance. It is worth noting that carboxylic acids are not terribly acidic. They are acidic when compared with other organic compounds, such as alcohols and amines, but not very acidic when compared with inorganic acids, such as sulfuric acid or nitric acid. In the equilibrium above showing a carboxylic acid losing a proton, we have one molecule losing its proton for every 10,000 molecules that do not give up their proton. In the world of acidity, this is not very acidic, but everything is relative.

So we have learned that resonance (which delocalizes a negative charge) is a stabilizing factor. The question now is how to roughly determine how stabilizing this factor is. Consider, for example, the following case:

The negative charge is stabilized over four atoms: one oxygen atom and three carbon atoms. Even though carbon is not as happy with a negative charge as oxygen is, nevertheless, it is better to spread the charge over one oxygen and three carbon atoms than to leave the negative charge stuck on one oxygen. Spreading the charge around helps to stabilize that charge.

But the number of atoms sharing the charge isn't everything. For example, it is better to have the charge spread over two oxygen atoms than to have the charge spread over one oxygen and three carbon atoms:

More stable

So now we have the basic framework to compare two compounds that are both resonance stabilized. We need to compare the compounds, keeping in mind the rules we just learned:

1. The more delocalized the better. A charge spread over four atoms gives a more stable compound than a charge spread over two atoms, *but*

2. One oxygen is better than many carbon atoms.

Now let's do some problems.

EXERCISE 3.6 Compare the two protons shown in the following compound. Which one is more acidic?

Answer We begin by pulling off one proton and drawing the conjugate base that we get. Then we do the same thing for the other proton:

Now we need to compare these conjugate bases and ask which one is more stable. In the compound on the left, we are looking at a charge that is localized on a nitrogen atom. For the compound on the right, the negative charge is delocalized over a nitrogen atom and an oxygen atom (draw resonance structures). It is more stable for the charge to be delocalized, so the second compound is more stable.

The more acidic proton is that one that leaves to give the more stable conjugate base.

PROBLEMS

3.7 Compare the two protons identified below. There are more protons in the compound, but only two of them are shown.

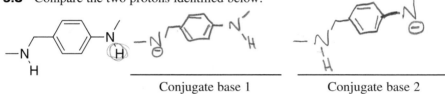

Conjugate base 1   Conjugate base 2

Identify which of these protons is more acidic, and explain why by comparing the stability of the conjugate bases.

_____

_____

3.8 Compare the two protons identified below:

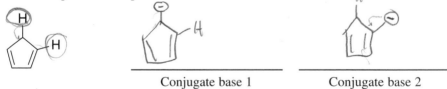

Conjugate base 1   Conjugate base 2

Identify which proton is more acidic, and explain why by comparing the stability of the conjugate bases.

_____

_____

3.9 Compare the two protons identified below:

Conjugate base 1   Conjugate base 2

Identify which proton is more acidic, and explain why by comparing the stability of the conjugate bases.

_____

_____

3.10 Compare the two protons identified below:

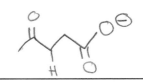

Conjugate base 1 _____   Conjugate base 2 _____

Identify which proton is more acidic, and explain why by comparing the stability of the conjugate bases.

_____

_____

3.11 Compare the two protons identified below:

Conjugate base 1 _____   Conjugate base 2 _____

Identify which proton is more acidic, and explain why by comparing the stability of the conjugate bases.

_____

_____

3.12 Compare the two protons identified below:

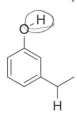
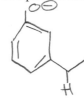
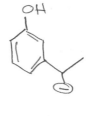

Conjugate base 1 _____   Conjugate base 2 _____

Identify which proton is more acidic, and explain why by comparing the stability of the conjugate bases.

_____

_____

3.3 FACTOR 3—INDUCTION

Let's compare the following compounds:

Which compound is more acidic? The only way to answer that question is to pull off the protons and draw the conjugate bases:

Let's go through the factors we learned so far. Factor 1 does not answer the problem: in both cases, the negative charge is on oxygen. Factor 2 also does not answer the problem: in both cases, there is resonance that delocalizes the charge over two oxygen atoms. Now we need factor 3.

The difference between the compounds is clearly the placement of the chlorine atoms. What effect will this have? For this, we need to understand a concept called induction.

We know that electronegativity measures the affinity of an atom for electrons, so what happens when you have two atoms of different electronegativity connected to each other? For example, consider a carbon–oxygen bond (C–O). Oxygen is more electronegative, so the two electrons that are shared between the carbon and oxygen (the two electrons that form the bond between them) are pulled more strongly by the oxygen atom. This creates a difference in the electron density on the two atoms—the oxygen becomes electron rich and the carbon becomes electron poor. This is usually shown with the symbols δ^+ and δ^-, which indicate "partial" positive and "partial" negative charges:

This "pulling" of electron density is called induction. Going back to our first example, the three chlorine atoms will pull electron density through induction from the carbon that they are attached to. This will make that carbon become electron poor (δ^+). This carbon can then pull electron density from the region that has the negative charge, and this effect will stabilize the negative charge:

More stable

Inductive effects fall off rapidly with distance, so there is a very large difference between the following compounds:

More stable

In fact, the second compound is not much different from the compound without the chlorine atoms entirely. So electronegative atoms that are too far away do not have much of an effect, but if they are too close, they can have a destabilizing effect:

In these compounds the protons identified are not very acidic, because they are too close to the electronegative atoms. If we deprotonate them, we will find a negative charge directly next to an atom with lone pairs. These lone pairs repel each other:

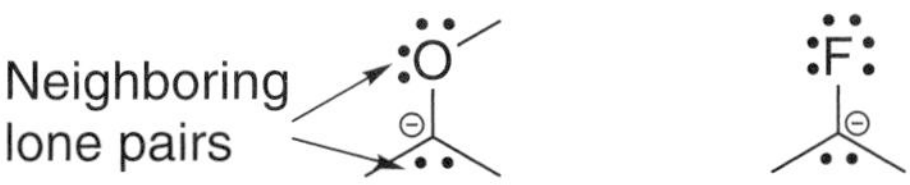

Neighboring lone pairs

When lone pairs exist on neighboring atoms, we call this the alpha effect. The lone pairs are repelling each other, and this is a destabilizing effect.

Now we know the effect that electronegative atoms (N, O, Cl, Br, etc.) can have if they are near (but not too close and not too far) from a negative charge. It is a stabilizing effect. But what effect do carbon atoms have (alkyl groups)? For example, is there a difference in acidity between the following two compounds?

Yes, there is, and it is very important to understand this difference, because the difference comes from a principle that will be applied in many different ways over the entire course. The idea is very simple: *alkyl groups are electron donating.*

This is so because of a concept called *hyperconjugation,* which we will not get into here; if you are interested, you can look it up in your textbook. But the bottom-line, take-home message is that alkyl groups are electron donating. So, what effect will this have on a negative charge? If electron density is being given to an area where there is a negative charge, then this area becomes less stable. It would be as if you are holding a hot potato, and someone with a hot iron heats up your potato even more.

So the comparison goes like this:

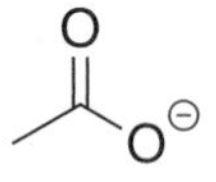

More stable

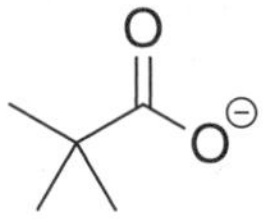

Less stable
Alkyl groups destabilize
this negative charge

and therefore,

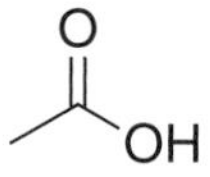

More acidic

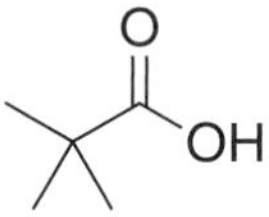

Less acidic

EXERCISE 3.13 Compare two protons shown in the following compound. Which proton is more acidic?

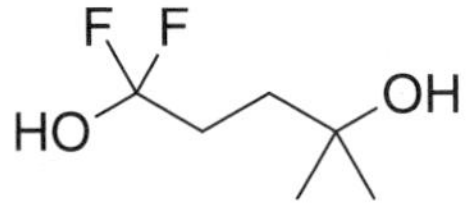

Answer Begin by drawing the conjugate bases:

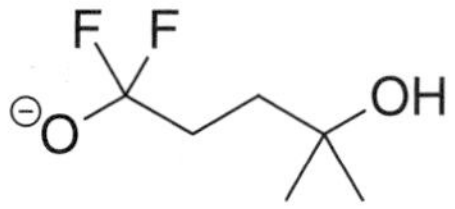

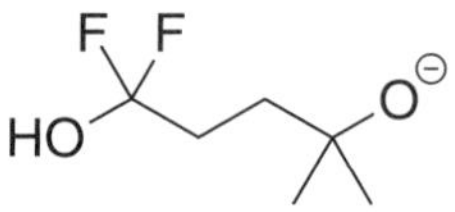

In the compound on the left, the charge is somewhat stabilized by the inductive effects of the two neighboring fluorine atoms. In contrast, the compound on the right is destabilized by the presence of two carbon atoms (methyl groups) that donate electron density. Therefore, the compound on the left is more stable.

The more acidic proton is the one that will leave to give the more stable negative charge. So the circled proton is more acidic:

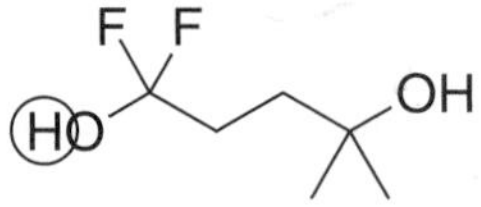

PROBLEMS

3.14 Compare the two protons identified below:

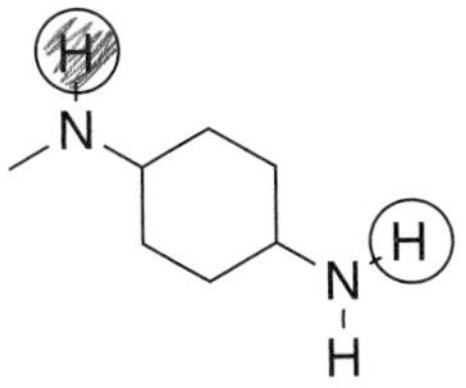

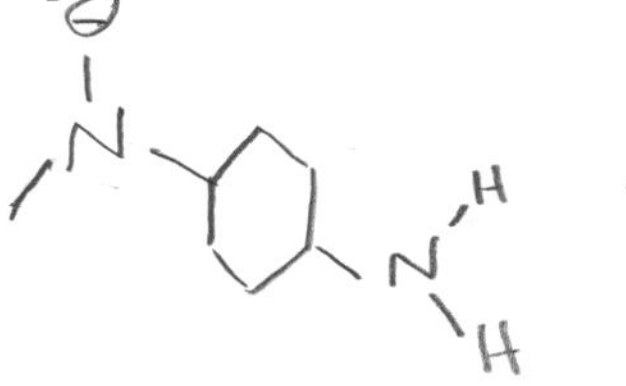

Conjugate base 1 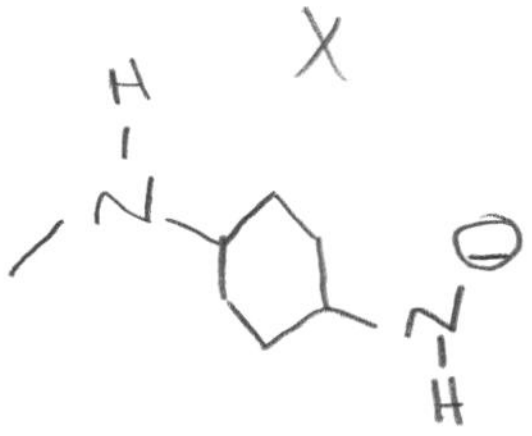   Conjugate base 2

Identify which proton is more acidic, and explain why by comparing the stability of the conjugate bases.

_____

_____

3.15 Compare the two protons identified below:

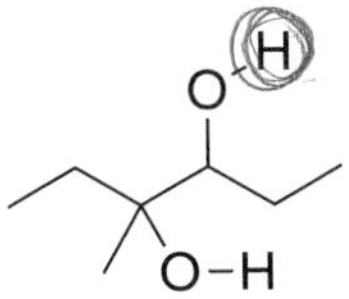

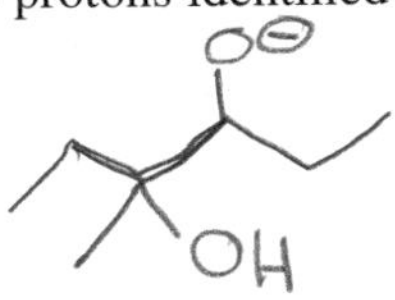

Conjugate base 1 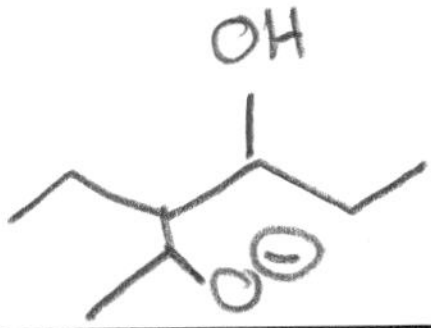   Conjugate base 2

Identify which proton is more acidic, and explain why by comparing the stability of the conjugate bases.

_____

_____

3.16 Compare the two protons identified below:

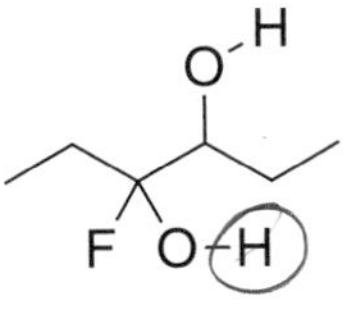

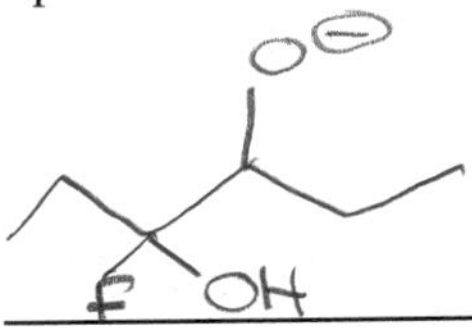

Conjugate base 1 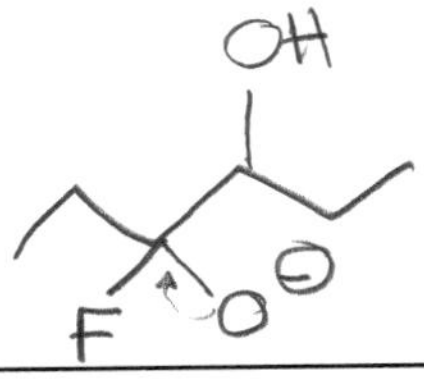   Conjugate base 2

Identify which proton is more acidic, and explain why by comparing the stability of the conjugate bases.

_____

_____

3.4   FACTOR 4—ORBITALS

The three factors we have learned so far will not explain the difference in acidity between the two identified protons in the compound below:

If we pull off the protons and look at the conjugate bases to compare them, we see this:

In both cases, the negative charge is on a carbon, so factor 1 does not help. In both cases, the charge is not stabilized by resonance, so factor 2 does not help. In both cases, there are no inductive effects to consider, so factor 3 does not help. The answer here comes from looking at the type of orbital that is accommodating the charge.

Let's quickly review the shape of hybridized orbitals. sp^3, sp^2, and sp orbitals all have roughly the same shape, but they are different in size:

sp^3 sp^2 sp

Notice that the sp orbital is smaller and tighter than the other orbitals. It is closer to the nucleus of the atom, which is located at the point where the front lobe (white) meets the back lobe (gray). Therefore, a lone pair of electrons residing in an sp orbital will be held closer to the positively charged nucleus and will be *stabilized* by being close to the nucleus.

So a negative charge on an sp hybridized carbon is more stable than a negative charge on an sp^3 or sp^2 hybridized carbon:

More stable

Determining which carbon atoms are sp, sp^2, or sp^3 is very simple: a carbon with a triple bond is sp, a carbon with a double bond is sp^2, and a carbon with all single bonds is sp^3. For more on this topic, turn to the next chapter (covering geometry).

EXERCISE 3.17   Locate the most acidic proton in the following compound:

Answer   It is important to recognize where all of the protons (hydrogen atoms) are. If you cannot do this, then you should review Chapter 1, which covers bond-line drawings. Only one proton can leave behind a negative charge in an sp orbital. All of the other protons would leave behind a negative charge on either sp^2 or sp^3 hybridized orbitals. So the most acidic proton is

3.5   RANKING THE FOUR FACTORS

Now that we have seen each of the four factors individually, we need to consider what order of importance to place them in. In other words, what should we look for first? And what should we do if two factors are competing with each other?

In general, the order of importance is the order in which the factors were presented in this chapter. Whenever you need to determine which proton is the most acidic, you need to compare all of the conjugate bases and ask which one is the most stable. To determine stability, here is what you should look for, in this order:

1. What atom is the charge on? (Remember the difference between comparing atoms in the same row and comparing atoms in the same column.)
2. Are there any resonance effects making one conjugate base more stable than the others?
3. Are there any inductive effects (electronegative atoms or alkyl groups) that stabilize or destabilize any of the conjugate bases?
4. In what orbital do we find the negative charge for each conjugate base that we are comparing?

There is an important exception to this order. Compare the two compounds below:

If we wanted to know which compound was more acidic, we would pull off the protons and compare the conjugate bases:

$$\text{———}\!\!\equiv\!\!\ominus \qquad\qquad\qquad ^{\ominus}\text{NH}_2$$

When comparing these two negative charges, we find two competing factors: the first factor (what atom is the charge on?) and the fourth factor (what orbital is the charge in?). The first factor says that a negative charge on nitrogen is more stable than a negative charge on carbon. However, the fourth factor says that a negative charge in an sp orbital is more stable than a negative charge in an sp^3 orbital (the negative charge on the nitrogen is an sp^3 orbital). In general, we would say that factor 1 wins over the others. But this case is an exception, and factor 4 (orbitals) actually wins here, so the negative charge on the carbon is more stable in this case:

$$\text{———}\!\!\equiv\!\!\ominus \qquad\qquad\qquad ^{\ominus}\text{NH}_2$$
More stable

In fact, for this reason, NH_2^- is generally used as the base for pulling the proton off of a triple bond.

There are, of course, other exceptions, but the one explained above is the most common. In most cases, you should be able to apply the four factors and provide a qualitative assessment of acidity.

EXERCISE 3.18 Compare the two protons identified below:

Identify which proton is more acidic, and explain why.

Answer The first thing we need to do is draw the conjugate bases:

Now we can compare them and ask which negative charge is more stable, using our four factors:

1. *Atom* In both cases, the charge is on an oxygen, so this doesn't help us.
2. *Resonance* The compound on the left has resonance stabilization and the compound on the right does not. Based on this factor alone, we would say the compound on the left is more stable.

3. *Induction* The compound on the right has an inductive effect that stabilizes the charge, but the compound on the left does not have this effect. Based on this factor alone, we would say the compound on the right is more stable.

4. *Orbital* This does not help us.

So, we have a competition of two factors. In general, resonance will beat induction, so we can say that the negative charge on the left is more stable. Therefore, the more acidic proton is the one circled here:

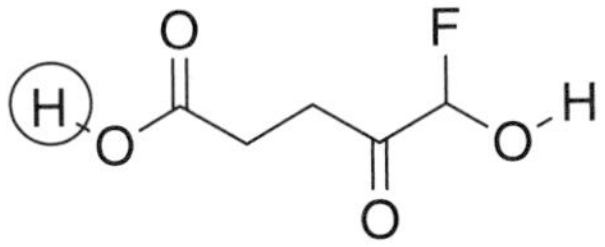

Remember the four factors, and what order they come in:

1. Atom
2. Resonance
3. Induction
4. Orbital

If you have trouble remembering the order, try remembering this acronym: ARIO.

PROBLEMS For each compound below, two protons have been identified. In each case, determine which of the two protons is more acidic.

3.19

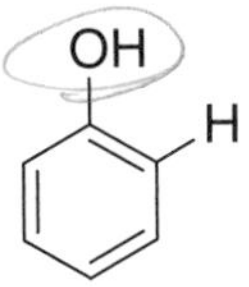

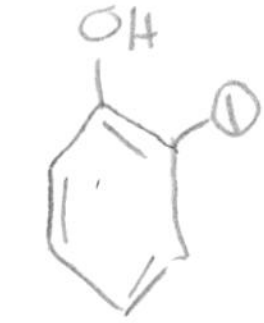

Conjugate base 1    Conjugate base 2

3.20

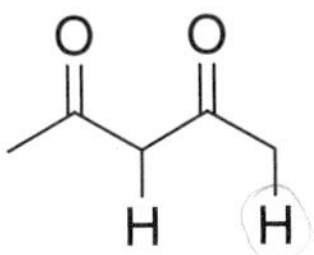

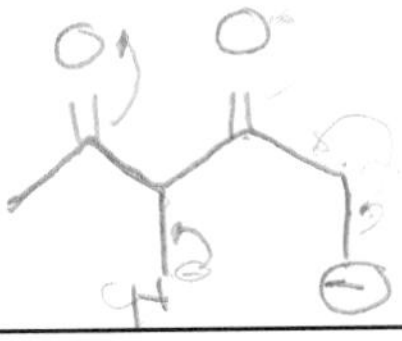

Conjugate base 1 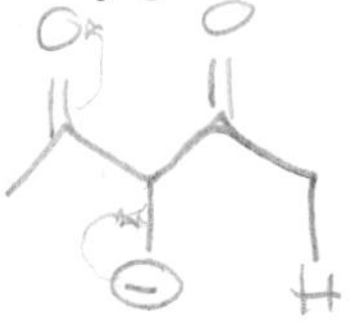   Conjugate base 2

3.21

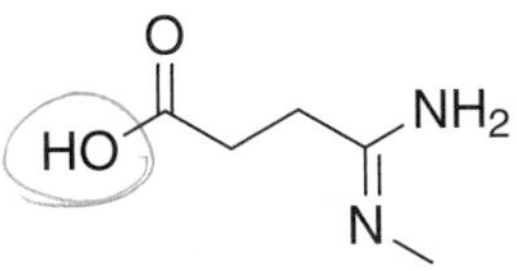

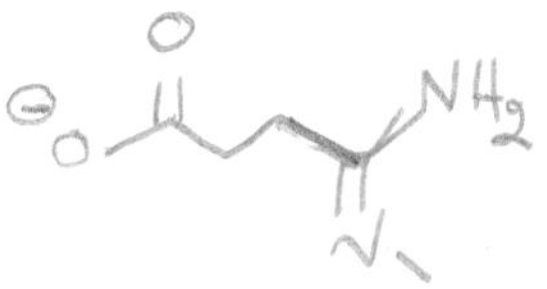

Conjugate base 1 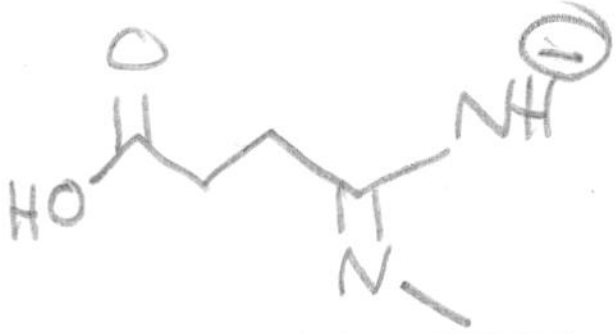   Conjugate base 2

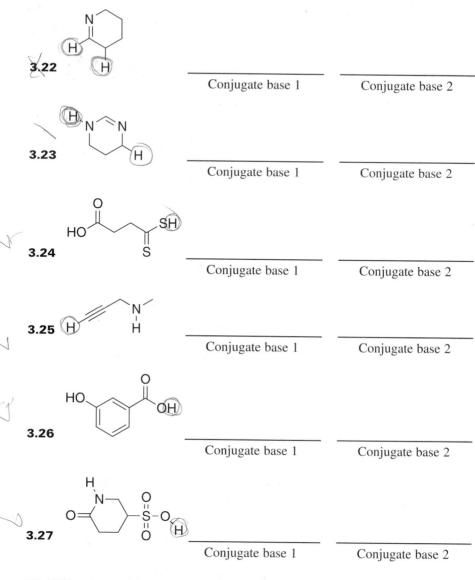

3.22

_____   _____
Conjugate base 1   Conjugate base 2

3.23

_____   _____
Conjugate base 1   Conjugate base 2

3.24

_____   _____
Conjugate base 1   Conjugate base 2

3.25

_____   _____
Conjugate base 1   Conjugate base 2

3.26

_____   _____
Conjugate base 1   Conjugate base 2

3.27

_____   _____
Conjugate base 1   Conjugate base 2

PROBLEMS For each pair of compounds below, predict which will be more acidic.

3.28 HCl   HBr   **3.29** H_2O H_2S

3.30 NH_3 CH_4 **3.31** H≡H   H   H

3.32   **3.33** Cl_3C CCl_3

3.6 QUANTITATIVE MEASUREMENT (pK_a VALUES)

Everything we have mentioned so far has been the *qualitative* method for comparing acidity of different protons. In other words, we never said how *much more* acidic one proton is over another, and we never said *exactly* how acidic each proton is. We have talked only about relative acidities: which proton is *more* acidic?

There is also a *quantitative* method of measuring acidities. All protons can be given a number that quantifies exactly how acidic they are. This value is called pK_a. It is impossible to figure out the exact pK_a by just looking at the structure of a compound. The pK_a must be determined empirically through experimentation. Many professors require that you know some general pK_a's for certain classes of compounds (for instance, all alcoholic protons, RO–H, will have the same ballpark pK_a). Most textbooks will have a chart that you can memorize. Your instructor will tell you if you are expected to memorize this chart. Either way, you should know what the numbers mean.

The smaller the pK_a, the more acidic the proton is. This probably seems strange, but that's the way it is. A compound with a pK_a of 4 is more acidic than a compound with a pK_a of 7. Next, we need to know what the difference is between 4 and 7. These numbers measure orders of magnitude. So the compound with a pK_a of 4 is 10^3 times more acidic (1000 times more acidic) than a compound with a pK_a of 7. If we compare a compound with a pK_a of 10 to a compound with a pK_a of 25, we find that the first compound is 10^{15} times more acidic than the second compound (1,000,000,000,000,000 times more acidic).

3.7 PREDICTING THE POSITION OF EQUILIBRIUM

Now that we know how to compare stability of charge, we can begin to predict which side of an equilibrium will be favored. Consider the following scenario:

$$\text{HA} \quad + \quad \text{B}^{\ominus} \quad \rightleftharpoons \quad \text{A}^{\ominus} \quad + \quad \text{HB}$$

This equilibrium represents the struggle between two compounds competing for H^+. A^- and B^- are competing with each other. Sometimes A^- gets the proton and sometimes B^- gets the proton. If we have a very large amount of A^- and B^- and not enough H^+ to protonate both of them, then at any given moment in time, there will be a certain number of A's that have a proton (HA) and a certain number of B's that have a proton (HB). These numbers are controlled by the equilibrium, which is controlled by (you guessed it) *stability of the negative charges*. If A^- is more stable than B^-, then A will be happy to have the negative charge and B^- will grab most of the protons. However, if B^- is more stable than A^-, then we will have the reverse effect.

Another way of looking at this is the following. In the equilibrium above, we see an A^- on one side and a B^- on the other side. The equilibrium will favor whichever side has the more stable negative charge. If A⁻ is more stable, then the equilibrium will lean so as to favor the formation of A⁻:

$$HA \quad + \quad \overset{\ominus}{B} \quad \longrightarrow \quad \overset{\ominus}{A} \quad + \quad HB$$

If B⁻ is more stable, then the equilibrium will lean so as to favor the formation of B⁻:

$$HA \quad + \quad \overset{\ominus}{B} \quad \rightleftharpoons \quad \overset{\ominus}{A} \quad + \quad HB$$

The position of equilibrium can be predicted once we know how to assess relative stability of negative charges.

EXERCISE 3.34　Predict the position of equilibrium for the following reaction:

$$H_2O \quad + \quad \overset{\ominus}{CH_3O} \quad \rightleftharpoons \quad \overset{\ominus}{HO} \quad + \quad CH_3OH$$

Answer　We look at both sides of the reaction and compare the negative charge on either side. Then we ask which one is more stable. We use the four factors:

1. *Atom*　The negative charge on the left is on oxygen, and negative charge on the right is also on oxygen. So this factor does not help us.
2. *Resonance*　Neither one is resonance stabilized.
3. *Induction*　The negative charge on the left is destabilized by an electron-donating alkyl group. The one on the right is not destabilized in this way.
4. *Orbital*　No difference between the right and left.

Based on factor 3, we conclude that the one on the right is more stable, and therefore the equilibrium lies to the right. We show this in the following way:

$$H_2O \quad + \quad \overset{\ominus}{CH_3O} \quad \rightleftharpoons \quad \overset{\ominus}{HO} \quad + \quad CH_3OH$$

PROBLEMS

3.35　Predict the position of equilibrium for the following reaction:

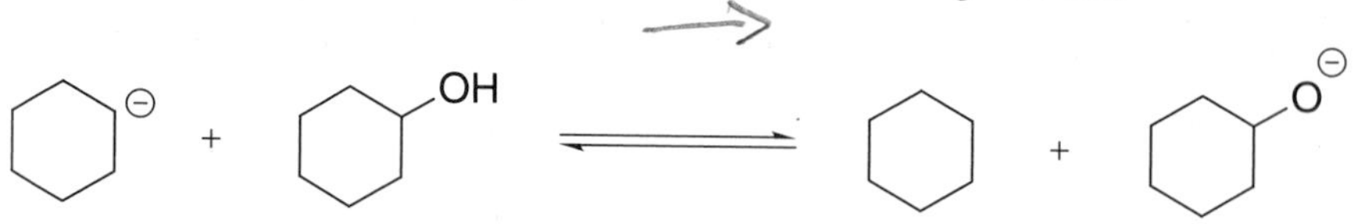

3.36　Predict the position of equilibrium for the following reaction:

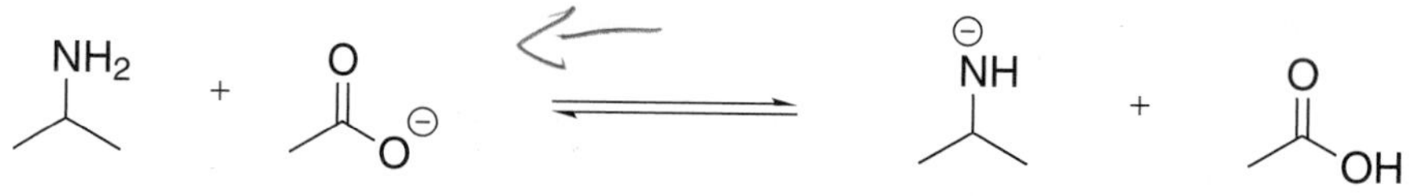

3.37 Predict the position of equilibrium for the following reaction:

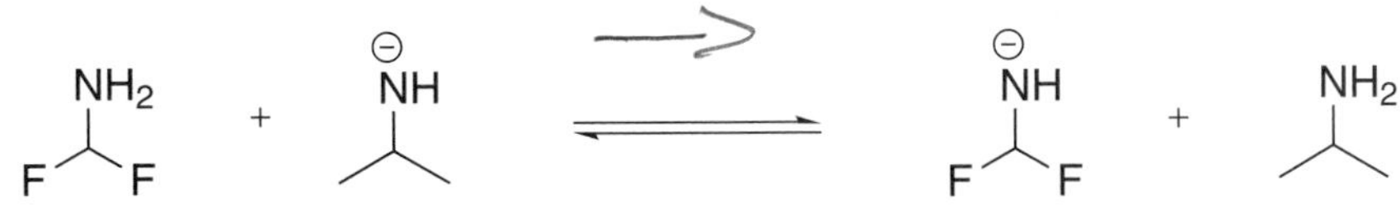

3.8 SHOWING A MECHANISM

Later on in the course, you will spend a lot of time drawing mechanisms of reactions. A mechanism shows how the electrons move during a reaction to form the products. Sometimes many steps are required, and sometimes just a few steps are required. In acid–base reactions, mechanisms are very straightforward because there is only one step. We use curved arrows (just like we did when drawing resonance structures) to show how the electrons flow. The only difference is that here we are allowed to break single bonds, because we are using arrows to show how a reaction happened (a reaction that involved the breaking of a single bond). With resonance drawings, we can never break a single bond (remember the first commandment). The second commandment—never violate the octet rule—is still true, even when we are drawing mechanisms. We can never violate the octet rule.

From an arrow-pushing point of view, all acid–base reactions are the same. It goes like this:

There are always two arrows. One is drawn coming from the compound with the negative charge and grabbing the proton. The second arrow is drawn coming from the bond (between the proton and whatever atom is connected to the proton) and going to the atom currently connected to the proton. That's it. There are always two arrows. Never 3 and never 1. Always 2. Each arrow has a head and a tail, so there are four possible mistakes you can make. You can get either of the heads wrong or either of tails wrong. With a little bit of practice you will see just how easy it is, and you will realize that acid–base reactions always follow the same mechanism.

EXERCISE 3.38 Show the mechanism for the following acid–base reaction:

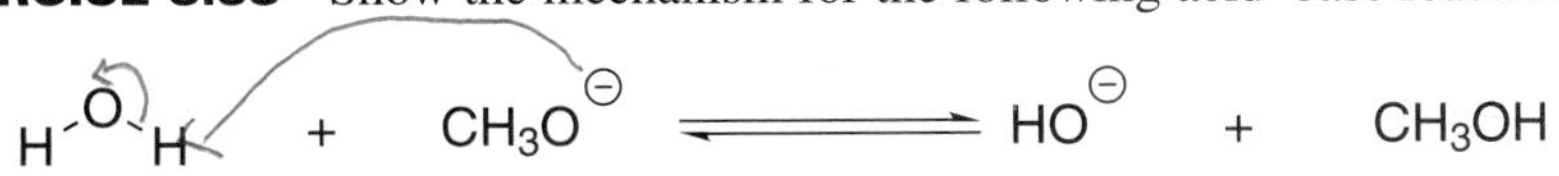

Answer Remember—2 arrows. One from the base to the proton and the other from the bond (that is losing the proton) to the atom (currently connected to the proton):

0115

PROBLEMS

3.39 Show the mechanism for the following acid–base reaction:

3.40 Show the mechanism for the following acid–base reaction:

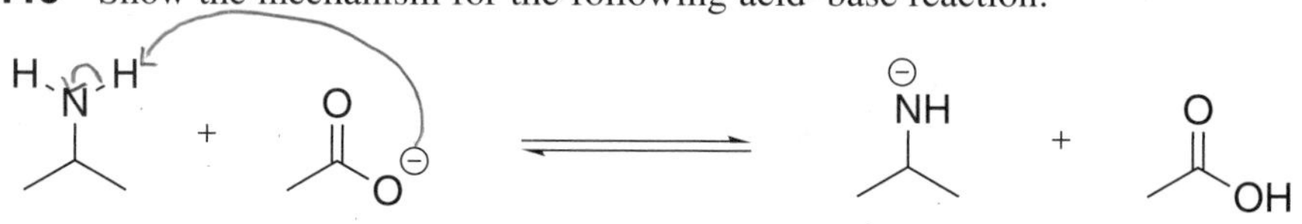

PROBLEMS Show the mechanism for the reaction that takes place when you mix hydroxide (HO⁻) with the each of following compounds (remember that you need to look for the most acidic proton in each case).

3.41

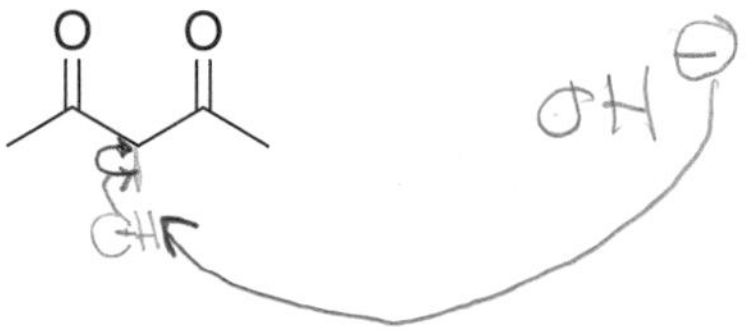

3.42

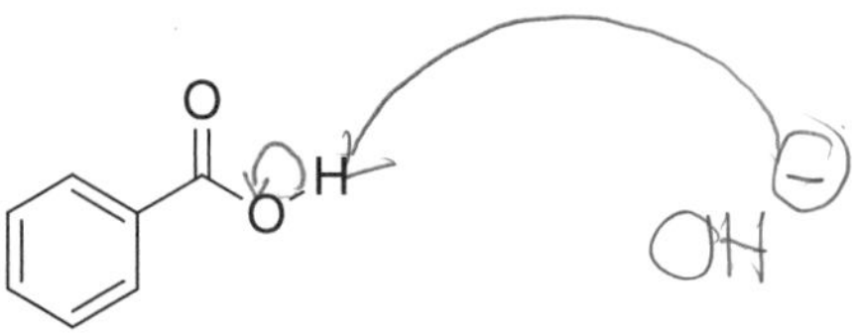

3.43

PROBLEMS   Show the mechanism for the reaction that takes place when you mix the amide ion (H_2N^-) with the each of following compounds (remember that you need to look for the most acidic proton in each case).

3.44   H————H

H_2N^-

3.45

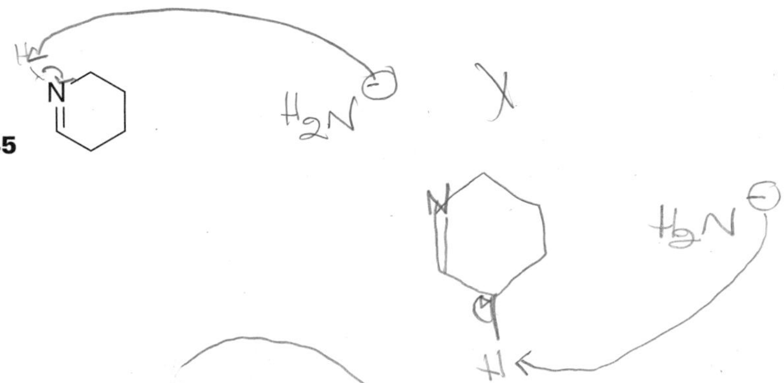

H_2N^-

H_2N^-

3.46

H_2N^-

H_2N^-

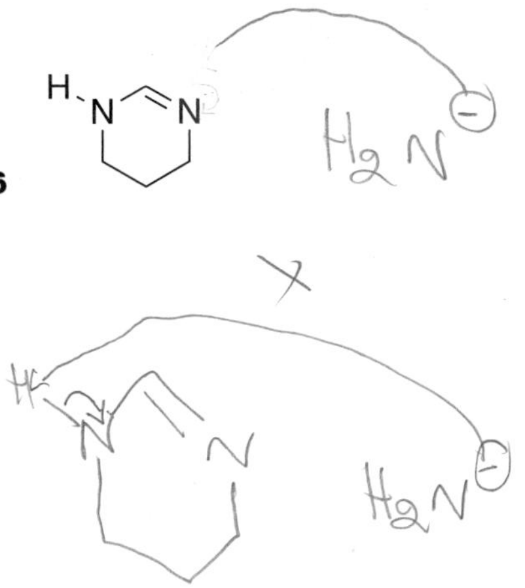

CHAPTER 4

GEOMETRY

In this chapter, we will see how to predict the 3D shape of molecules. This is important because it limits much of the reactivity that you will see in the second half of this course. For molecules to react with each other, the parts of the molecules that can react with each other must be able to get close in space. If the geometry of the molecules prevents them from getting close, then there cannot be a reaction. This concept is called *sterics.*

Let's use an analogy to help us see the importance of geometry. Imagine that you are stuffing a turkey for Thanksgiving dinner and your hand gets stuck inside the turkey. Just at that moment, someone wants to shake your hand. You can't shake the person's hand because your hand is unavailable at the moment. It's kind of the same way with molecules. When two molecules react with each other, there are specific sites on the molecules that are reacting with each other. If those sites cannot get close to each other, the reaction won't happen.

There will be many times in the second half of this course when you will be trying to determine which way a reaction will proceed from two possible outcomes. Many times, you will choose one outcome, because the other outcome has steric problems to overcome (the geometry of the molecules does not permit the reactive sites to get close together). In fact, you will learn to make decisions like this as soon as you learn your first reactions: S_N2 versus S_N1 reactions. Now that we know why geometry is so important, we need to brush up on some basic concepts.

To determine the geometry of an entire molecule, we need to be able to determine the geometry of each atom based on how it is connected to the atoms around it. After all, that is what determines the geometry—how the atoms are connected in 3D space. Since atoms are connected to each other with bonds, it makes sense that we need to take a close look at bonds. In particular, we need to know the exact locations and angles of every bond to every atom. This might sound difficult, but it is actually straightforward, and with a little bit of practice, you can get to the point where you know the geometry of a molecule as soon as you look at it (without even needing to think about it). That is the point that we need to get to, and that is what this chapter is all about.

4.1 ORBITALS AND HYBRIDIZATION STATES

To determine the geometry of a molecule, we need to know how atoms bond with each other three dimensionally, so it makes sense for our discussion to start with orbitals. After all, bonds come from overlapping orbitals.

A bond is formed when an electron of one atom overlaps with an electron of another atom. The two electrons are shared between both atoms, and we call that a bond. Since electrons exist in regions of space called orbitals, then what we really need to know is; what are the locations and angles of the orbitals around every atom? It is not so complicated, because the number of possible arrangements of orbitals is very small. You need to learn the possibilities, and how to identify them when you see them. So, we need to talk about orbitals.

There are two simple orbitals: s and p orbitals (we don't really deal with d and f orbitals in organic chemistry). s orbitals are spherical and p orbitals have two lobes (one front lobe and one back lobe):

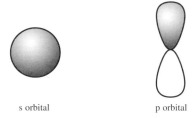

s orbital   p orbital

Atoms in the second row (such as C, N, O, and F) have one s orbital and three p orbitals in the valence shell. These orbitals are usually mixed together to give us hybridized orbitals (sp^3, sp^2, and sp). We get these orbitals by mixing the *properties* of s and p orbitals. What do we mean by mixing?

Imagine one swimming pool shaped like a triangle and another shaped like a pentagon; now we put them next to each other. We wave a magic wand and they magically turn into two rectangular pools. That would be a neat trick. That's what sp orbitals are: we take one s orbital and one p orbital, then wave a magic wand, and poof—we now have two equivalent orbitals that look the same. The two new orbital have a different shape from the original two orbitals. This new shape is somewhat of an average of the two original shapes.

If we mix two p orbitals and one s orbital, then we get three equivalent sp^2 orbitals. Let's go back to the pool analogy. Imagine two pools shaped like octagons and one shaped like a triangle. We wave our magic wand and get three pools shaped like hexagons. We started with three pools and we ended with three pools. But the three pools in the end are all the same and their shape is an average of the shapes of the original three pools. The same thing is true here with orbitals. We start with three orbitals (two p orbitals and one s orbital). Then we mix them together and end up with three orbitals that all look the same. All three new orbitals have the "average" properties of the original three orbitals. The three new orbitals (since they came from one s orbital and two p orbitals) are called sp^2 orbitals. Similarly, when you combine three p orbitals and one s orbital, you get four equivalent sp^3 orbitals.

To truly understand the geometry of bonds, we need to understand the geometry of these three different hybridization states. The hybridization state of an atom describes the type of hybridized atomic orbitals (sp^3, sp^2, or sp) that contain the valence electrons. Each hybridized orbital can be used either to form a bond with another atom or to hold a lone pair.

It is not difficult to determine hybridization states. If you can add, then you should have no trouble determining the hybridization state of an atom. Just count how many other atoms are bonded to your atom, and count how many lone pairs your atom has. Add these numbers. Now you have the total number of hybridized orbitals that contain the valence electrons. This number is all you need to determine the hybridization state of the atom. That probably sounded complicated, so let's look at an example to clear it up.

Consider the molecule below:

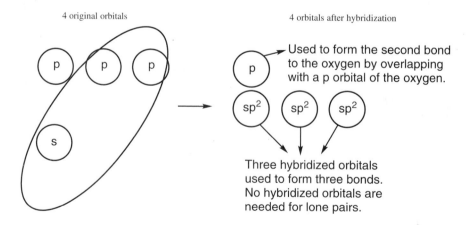

Let's try to determine the hybridization state of the carbon in the center. We begin by counting the number of atoms connected to this carbon atom. There are 3 atoms (O, H, and H). *The oxygen atom only counts as one.*

Next we count the number of lone pairs on the carbon atom. There are no lone pairs on the carbon atom. (If you are not sure how to tell that there are no lone pairs there, go back to Chapter 1 and review the section on counting lone pairs.) Now we take the sum of the attached atoms and the number of lone pairs—in this case, $3 + 0 = 3$. Therefore, three hybridized orbitals are being used here. That means that we have mixed two p orbitals and one s orbital (a total of three orbitals) to get three equivalent sp^2 orbitals. Thus, the hybridization is sp^2. Let's take a closer look at how this works.

Recall that the second row elements have three p orbitals and one s orbital that can be hybridized in one of three ways: sp^3, sp^2, or sp. If we are using three hybridized orbitals, then we must have mixed two p orbitals with one s orbital:

So here's the rule: Just add the number of bonded atoms to the number of lone pairs. The number you get tells you how many hybridized orbitals you need according to the following:

If the sum is 4, then you have 4 sp³ orbitals

If the sum is 3, then you have 3 sp² orbitals and one p orbital (as in our example)

If the sum is 2, then you have 2 sp orbitals and two p orbitals

---

There is one exception, which you will see in the chapter on aromaticity in your textbook. For now, let's not worry about it.

Once you get used to looking at drawings of molecules, you should not have to count anymore. There are certain arrangements that are always sp³ hybridized, and the same is true for sp² and sp. Here are some common examples:

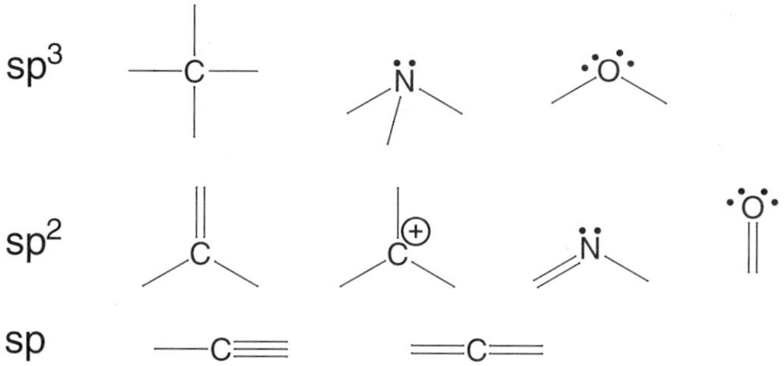

If you can determine the hybridization state of any atom, you will be able to easily determine the geometry of that atom. Let's do another example.

EXERCISE 4.1 Identify the hybridization state for the nitrogen atom in ammonia (NH₃).

Answer First we need to ask how many atoms are connected to this nitrogen atom. There are three hydrogen atoms. Next we need to ask how many lone pairs the nitrogen atom has. It has 1 lone pair. Now, we take the sum. 3 + 1 = 4. If we need to have four hybridized orbitals, then the hybridization state must be sp³.

PROBLEMS For each compound below, identify the hybridization state for the central carbon atom.

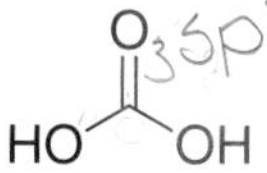

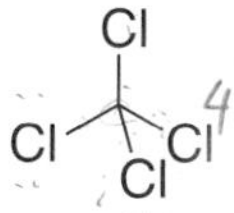

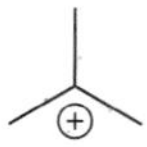

4.8   For each carbon atom in the following molecule, identify the hybridization state. Do not forget to count the hydrogen atoms (they are not shown). Use the following simple method: A carbon with 4 single bonds is sp^3 hybridized. A carbon with a double bond is sp^2 hybridized, and a carbon with a triple bond is sp hybridized.

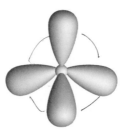

Once you get used to it, you do not need to count anymore—just look at the number of bonds. If carbon has only single bonds, then it is sp^3 hybridized. If the carbon atom has a double bond, then it is sp^2 hybridized. If the carbon atom has a triple bond, then it is sp hybridized. Consult the chart of common examples on the previous page.

4.2   GEOMETRY

Now that we know how to determine hybridization states, we need to know the geometry of each of the three hybridization states. One simple theory explains it all. This theory is called the *valence shell electron pair repulsion theory* (VSEPR). Stated simply, all orbitals containing electrons in the outermost shell (the valence shell) want to get as far apart from each other as possible. This one simple idea is all you need to predict the geometry around an atom. First, let's apply the theory to the three types of hybridized orbitals.

   1. Four equivalent sp^3-hybridized orbitals achieve maximum distance from one another when they arrange in a tetrahedral structure:

Think of this as a tripod with an additional leg sticking straight up in the air. In this arrangement, each of the four orbitals is exactly 109.5° from each of the other three orbitals.

2. Three equivalent sp²-hybridized orbitals achieve maximum distance from one another when they arrange in a trigonal planar structure:

sp²

All three orbitals are in the same plane, and each one is 120° from each of the other orbitals. The remaining p orbital is orthogonal to (perpendicular to the plane of) the three hybridized orbitals.

3. Two equivalent sp-hybridized orbitals achieve maximum distance from one another when they arrange in a linear structure:

Both orbitals are 180° from each other. The remaining two p orbitals are 90° from each other and from each of the hybridized orbitals.

So far its very simple:

1. sp³ = tetrahedral
2. sp² = trigonal planar
3. sp = linear

But here's where students usually get confused. What happens when a hybridized orbital holds a lone pair? What does that do to the geometry? The answer is that the geometry of the orbitals does not change, but the geometry of the molecule is affected. Why?

Let's look at an example. In ammonia (NH₃), the nitrogen atom is sp³ hybridized, so *all four orbitals arrange in a tetrahedral structure,* just as we would expect. But only three of the orbitals in this arrangement are responsible for bonds. So, if we look just at the atoms that are connected, we do not see a tetrahedron. Rather, we see a trigonal pyramidal arrangement:

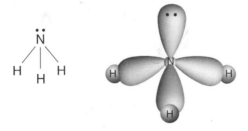

Trigonal, because there are three bonds pointing away from the central nitrogen atom, and pyramidal because it's shaped like a pyramid.

Similarly, in H_2O, the oxygen is sp^3 hybridized. So all four orbitals are in a tetrahedral arrangement, just as we would expect for an sp^3 hybridized atom. But only two of the orbitals are being used for bonds. So if we look just at the atoms that are connected, we do not see a tetrahedron. Rather, we see a bent arrangement:

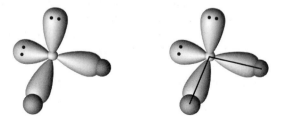

Let's now put all of this information together:

$$
\begin{array}{lll}
sp^3 & \text{with} & \text{0 lone pairs} = \text{tetrahedral} \\
sp^3 & \text{with} & \text{1 lone pair} = \text{trigonal pyramidal} \\
sp^3 & \text{with} & \text{2 lone pairs} = \text{bent} \\
sp^2 & \text{with} & \text{0 lone pairs} = \text{trigonal planar} \\
sp^2 & \text{with} & \text{1 lone pair} = \text{bent} \\
sp & \text{with} & \text{0 lone pairs} = \text{linear}
\end{array}
$$

That's it. There are only six different types of geometry that we need to know. First we determine the hybridization state. Then, using the number of lone pairs, we can figure out which of the six different types of geometry we are dealing with. Let's try it out on a problem.

EXERCISE 4.9   Identify the geometry of the carbon atom below:

$$
\begin{array}{c}
O \\
\parallel \\
H \diagdown C \diagup H
\end{array}
$$

Answer   First, we need to determine the hybridization state. We did this for this molecule earlier in this chapter and found that the hybridization state is sp^2 (there are 3 atoms connected and no lone pairs, so we need three hybridized orbitals; therefore, it is sp^2).

Next we remind ourselves how many lone pairs there are; in this case, there are none. So the geometry must be trigonal planar.

Once you can determine the geometry around an atom, you should have no problem determining the geometry, or shape, of a molecule. Simply repeat your analysis for each and every atom in the molecule. This may seem like a large task at first, but once you get the hang of it, you will be able to tell the geometry of an atom immediately upon seeing it.

For the next set of problems, you should get to the point where you can do these problems very quickly. The first few will take you longer than the last ones. If the last problem is still taking you a long time, then you have not mastered the process and you will need more practice. If this is the case, open to any page in the second half of your textbook. You will probably see drawings of structures. Point to any atom in a structure and try to determine what the geometry is. Use the list above to help you. Go from one drawing to the next until you can do it without the list. That is the important part—doing it without needing the list.

PROBLEMS For each compound below, identify the hybridization state and geometry for every atom in the compound.

4.10

4.11

4.12

4.13

4.14

4.15

4.16

4.17

NOMENCLATURE

All molecules have names, and we need to know their names to communicate. Consider the molecule below:

Clearly, it would be inadequate to refer to this compound as "you know, that thing with five carbons and an OH coming off the side with a chlorine on a double bond." First of all, there are too many other compounds that fit that fuzzy description. And even if we could come up with a very adequate description that could only be this one compound, it would take way too long (probably an entire paragraph) to describe. By following the rules of nomenclature, we can unambiguously describe this molecule with just a few letters and numbers: Z-2-chloropent-2-en-1-ol.

It would be impossible to memorize the names of every molecule, because there are too many to even count. It would also be impossible to memorize the name of every compound. Instead, we have a very systematic way of naming molecules. What you need to learn are the rules for how to give a name to a molecule (these rules are referred to as IUPAC nomenclature). This is a much more manageable task than memorizing names, but even these rules can become challenging to master. There are so many of them, that you could study only these rules for an entire semester and still not finish all of them. The larger the molecules get, the more rules you need to account for every kind of possibility. In fact, the list of rules is regularly updated and refined.

Fortunately, you do not need to learn all of these rules, because we deal with very simple molecules in this course. You need to learn only the rules that allow you to name small molecules. This chapter focuses on most of the rules you need to name simple molecules.

There are five parts to every name:

Stereoisomerism	Substituents	Parent	Unsaturation	Functional group

1. *Stereoisomerism*   Indicates whether double bonds are cis/trans or E/Z, and indicates stereocenters (R, S), which we will cover in the chapter on configuration.

2. *Substituents*   Are groups coming off of the main chain.

3. *Parent*   Is the main chain.

4. *Unsaturation* Identifies if there are any double or triple bonds.

5. *Functional group* The functional group after which the compound is named.

Let's use the compound above as an example:

Stereoisomerism	Substituents	Parent	Unsaturation	Functional group
Z	2-chloro	pent	2-en	1-ol

We will systematically go through all five parts to every name, starting at the end (functional group) and working our way backward to the first part of the name (stereoisomerism). It is important to do it backward like this, because the position of the functional group affects which parent chain you choose.

5.1 FUNCTIONAL GROUP

Stereoisomerism	Substituents	Parent	Unsaturation	Functional group

The term *functional group* refers to specific arrangements of atoms that have certain characteristics for reactivity. For example, when an –OH is connected to a compound, we call the molecule an alcohol. Alcohols will display similar reactivity, because alcohols all have the same functional group, the –OH group. In fact, most textbooks have chapters arranged according to functional groups (one chapter on alcohols, one chapter on amines, etc.). Accordingly, many textbooks treat nomenclature as an ongoing learning process: As you work through the course, you slowly add to your list of functional group names. Here we focus on six common functional groups, because you will certainly learn at least these six throughout your course.

When a compound has one of these six groups, we show it in the name of the compound by placing a suffix on the name of the molecule. As we saw, this is the last part of any name. So we need to know the suffixes that we use for each of these groups:

Functional group	Class of compound	Suffix
O ‖ R—C—OH	Carboxylic acid	-oic acid
O ‖ R—C—OR	Ester	-oate

(continued)

Functional group	Class of compound	Suffix
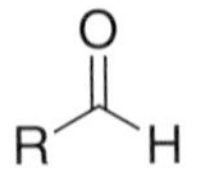 O ‖ R—H	Aldehyde	-al
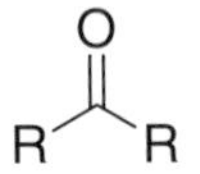 O ‖ R—R	Ketone	-one
R—O—H	Alcohol	-ol
H | R—N—H	Amine	-amine

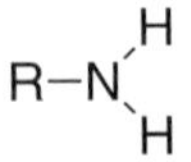

Halogens (F, Cl, Br, I) are usually not named in the suffix of a compound. They get named as substituents, which we will see later on.

Notice that the carboxylic acid is like a ketone and an alcohol placed next to each other. But be careful, because carboxylic acids are very different from ketones or alcohols. So don't make the mistake of thinking that a carboxylic acid is a ketone and an alcohol:

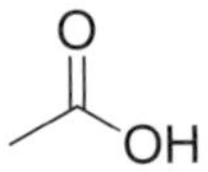

A carboxylic acid

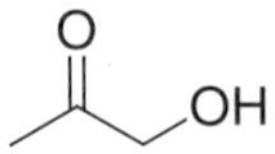

A ketone and an alcohol

The compound above that is a ketone and an alcohol brings about an important issue: how do you name the functional group when you have two functional groups in a compound? One will go in the suffix of the name and the other will be a prefix in the substituent part of the name. But how do we choose which one goes as the suffix of the name? There is a hierarchy that needs to be followed. The six groups shown above are listed according to their hierarchy, so a carboxylic acid takes precedence over an alcohol. A compound with both of these groups is named as an -oic acid, and we put the term "hydroxy" in the substituent part of the name.

EXERCISE 5.1   Identify what suffix you would use in naming the following compound:

HO⌒⌒NH₂

Answer There are two functional groups in this compound, so we have to decide between calling this compound an amine or calling it an alcohol. If we look at the hierarchy above, we see that an alcohol outranks an amine. Therefore, we use the suffix -ol in naming this compound.

PROBLEMS Identify what suffix you would use in naming each of the following compounds.

5.2 Suffix: <u>one</u>

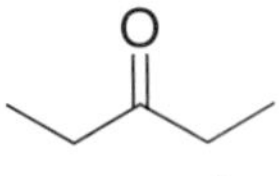

5.3 Suffix: <u>oate</u>

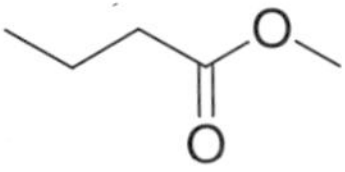

5.4 Suffix: <u>al</u>

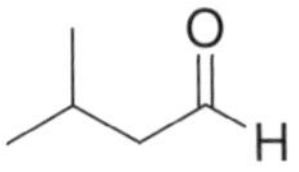

5.5 H₂N ⟋⟍ Cl Suffix: <u>amine</u>

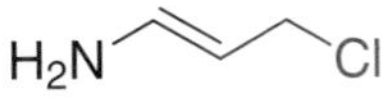

5.6 Br / OH Suffix: <u>ol</u>

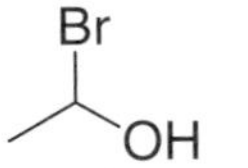

5.7 F₃C–CF₂–OH Suffix: <u>ol</u>

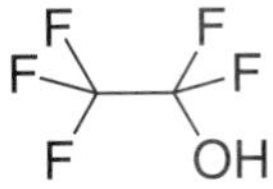

5.8 HO Br Suffix: <u>al</u>

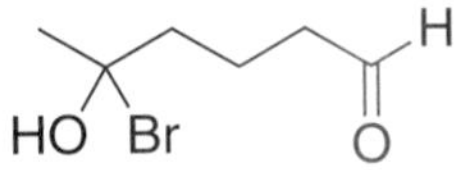

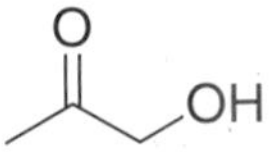

5.9 Suffix: <u>carboxylic</u> <u>one</u> <u>acid</u>

5.10 Suffix: <u>hydroxi</u> <u>oic acid</u>

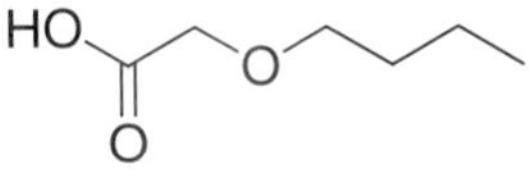

If there is no functional group in the compound, then we put an "e" at the end of the name:

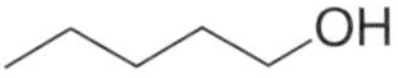

pentan**ol**

pentan**e**

5.2 UNSATURATION

Stereoisomerism	Substituents	Parent	Unsaturation	Functional group

Many molecules have double or triple bonds, often called "unsaturation" because a compound with a double or triple bond has less hydrogen than it would have without the double or triple bond. These double and triple bonds are very easy to see in bond-line drawings:

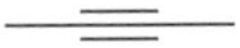

Triple bond

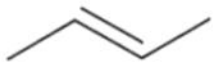

Double bond

Look above at the example of pentane. The "e" told us that there was no functional group. Working our way backward through the name, the "an" tells us that there are no double or triple bonds in the molecule. Double bonds are called "en" (pronounced *een*) and triple bonds are called "yn" (pronounced *ine*). For example,

pent<u>ene</u>

pent<u>yne</u>

If there are two double bonds in a compound, then the unsaturation is "-dien-". Three double bonds is "-trien-". Similarly, two triple bonds is "-diyn-", and three triple bonds is "-triyn-". For multiple double and triple bonds, we use the following terms:

di = 2   penta = 5

tri = 3   hexa = 6

tetra = 4

You will rarely ever see this many double or triple bonds in one compound, but it is possible to see double and triple bonds in the same molecule. For example,

The compound shown here has three double bonds and two triple bonds. So it is a triendiyne. Double bonds always get listed first.

EXERCISE 5.11   Identify how you would describe the unsaturation in the name of the following compound:

Answer   This compound has one double bond and one triple bond. For the double bond, we use the term "en". For the triple bond, we use the term "yn". Double bonds get listed first, so this compound is –enyn-.

PROBLEMS   Identify how you would describe the unsaturation in the name of the following compounds.

5.12
_- en -_

5.13
_- yn -_

5.14
dien

5.15
trien

5.16
trien

5.17
endiyn

5.3 NAMING THE PARENT CHAIN

Stereoisomerism	Substituents	Parent	Unsaturation	Functional group

When naming the parent of the compound, we are looking for the chain of carbon atoms that is going to be the root of our name. Everything else in the compound is connected to that chain at a specific location, designated by numbers. So we need to know how to choose the parent carbon chain and number it correctly.

The first step is learning how to say "a chain of three carbons" or "a chain of seven carbons." Here is a table showing the appropriate names:

Number of carbon atoms in the chain	Parent
1	meth
2	eth
3	prop
4	but
5	pent
6	hex
7	hept
8	oct
9	non
10	dec

If we have carbon atoms in a ring, we add the term cyclo, so a ring of six carbon atoms is called cyclohex- as the parent and a ring of five carbon atoms is cyclopent-.

You must commit these terms to memory. I am not a big advocate of memorization, but for now, you must memorize these terms. After a while, it will become habitual, like a phone number that you dial all of the time, and you won't have to think about it anymore.

The tricky part comes when you need to figure out which carbon chain to use. Consider the following example, which has three different possibilities for the parent chain:

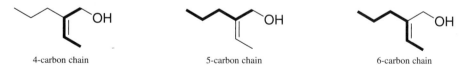

4-carbon chain   5-carbon chain   6-carbon chain

So how do we know whether to call this –but- (which is 4) or –pent- (which is 5) or –hex- (which is 6)? There is a hierarchy for this as well. The chain should be as long as possible, making sure to include the following groups, in this order:

Functional group

Double bond

Triple bond

First we need to find the functional group and make sure that the functional group is in the chain. Remember from last section that if there are two functional groups, one of them gets priority. The functional group that gets priority is the one that needs to be in our parent chain. Of the three possibilities shown above, this rule eliminates the last possibility, because it does not include the functional group on the parent chain. If there are still more choices of possible parent chains (as there are in this case), then we look for the chain that also includes the double bond (if there is one in the compound). In our case, there is a double bond, and this rule determines for us which chain to use:

Of the three possibilities, this is the only parent chain containing the functional group and the double bond. "Containing the functional group" means that the OH is connected to a carbon that is part of the chain. We do not count the oxygen itself as part of the chain. It is only attached to the chain. So the chain above is made up of four carbon atoms.

In cases where there is no functional group, then we look for the longest chain that includes the double bond. If there is no double bond, then we look for a triple bond, and choose the longest chain that has the triple bond in it.

If there are no functional groups, no double bonds, and no triple bonds, then we simply choose the longest chain possible.

Now you can see why we are moving our way backward through the naming process. We cannot name the parent correctly unless we can pick out the highest ranking functional group in the compound. So we start naming a compound by first asking which functional group has priority.

EXERCISE 5.18 Name the parent chain in the following compound:

Answer First we look for a functional group. There is only one, so we know the parent chain must include the carboxylic acid group. Next we look for a double bond. The parent chain should include that as well. This gives us our answer. The triple bond will not be included in the parent chain, because the functional group and the double bond are higher priority than a triple bond.

So we count the number of carbons in this chain. There are six (notice that we include the carbon of the carboxylic acid group). Therefore, the parent will be called
"hex".

PROBLEMS   Name the parent chain in the each of the following compounds.

5.19 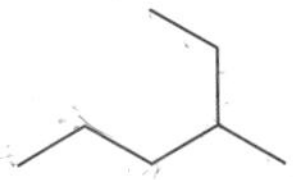   Parent: _hex_

5.20 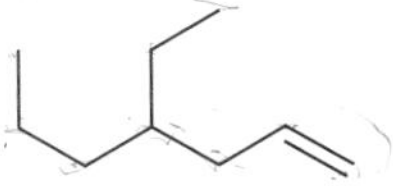   Parent: _Hept_

5.21 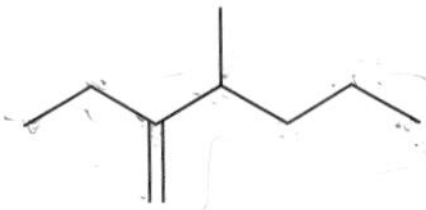   Parent: _hex_

5.22 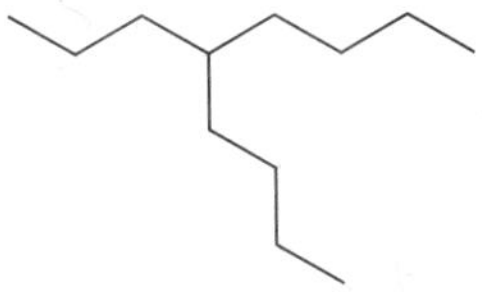   Parent: _non_

5.23 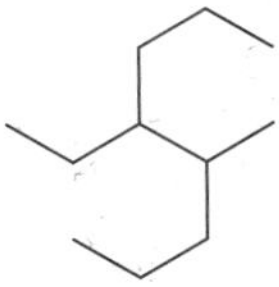   Parent: _oct_

5.24 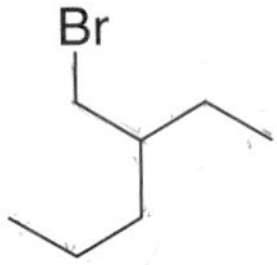   Br   Parent: _hex_

5.25 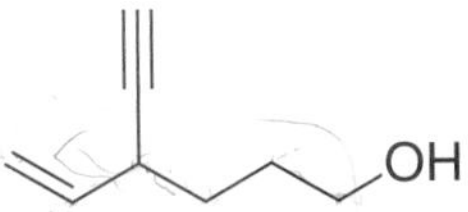   OH   Parent: _hex_

5.26 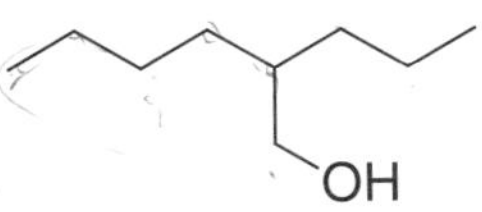   OH   Parent: _hexre_

5.27 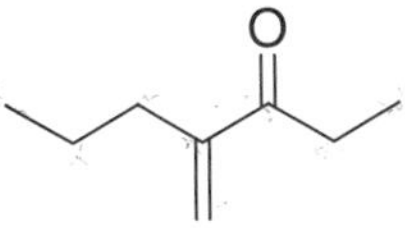   O   Parent: _pent_

5.4   NAMING SUBSTITUENTS

Stereoisomerism	Substituents	Parent	Unsaturation	Functional group

Once we have identified the functional group and the parent chain, then everything
else connected to the parent chain is called a substituent. In the following example,
all of the circled groups are substituents, because they are not part of the parent
chain:

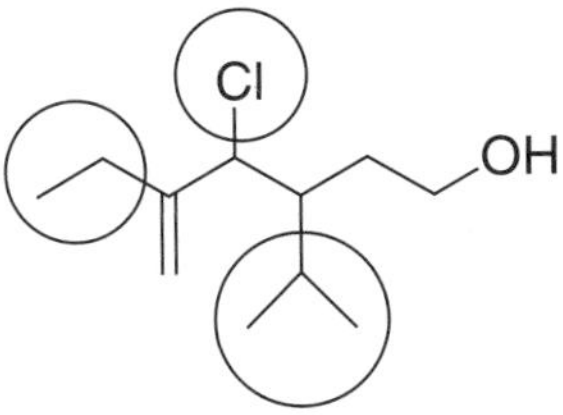

We start by learning how to name the alkyl substituents. These groups are named with the same terminology that parent chains are named, but we add "yl" to the end to imply that it is a substituent:

Number of carbon atoms in the substituent	Substituent
1	methyl
2	ethyl
3	propyl
4	butyl
5	pentyl
6	hexyl
7	heptyl
8	octyl
9	nonyl
10	decyl

Methyl groups can be shown in a number of ways, and all of them are acceptable:

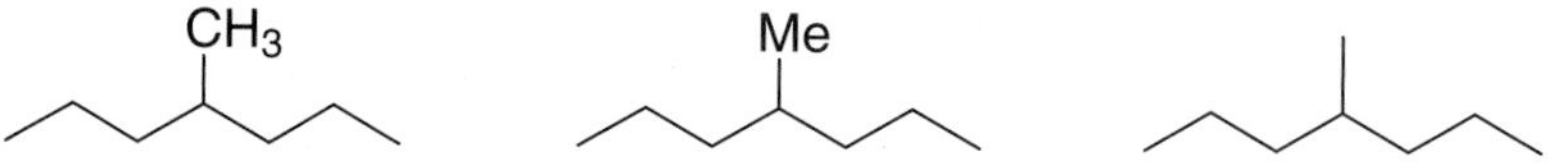

Ethyl groups can also be shown in a number of ways:

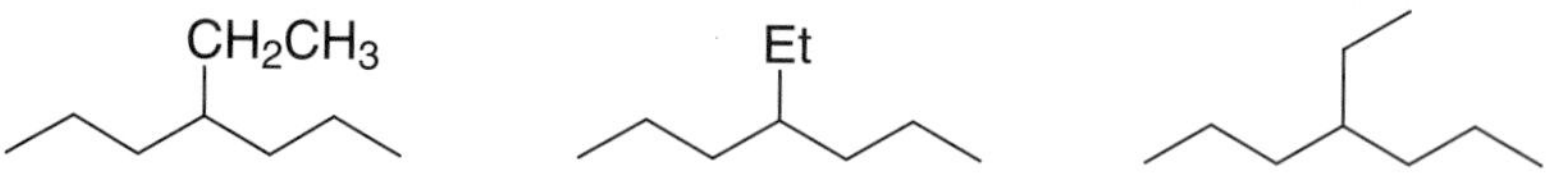

Propyl groups are usually just drawn, but sometimes you will see the term Pr (which stands for propyl):

Look at the propyl group above and you will notice that it is a small chain of 3 carbon atoms that is attached to the parent chain by the first carbon of the small chain. But what if it is attached by the middle carbon? Then it is not called propyl anymore:

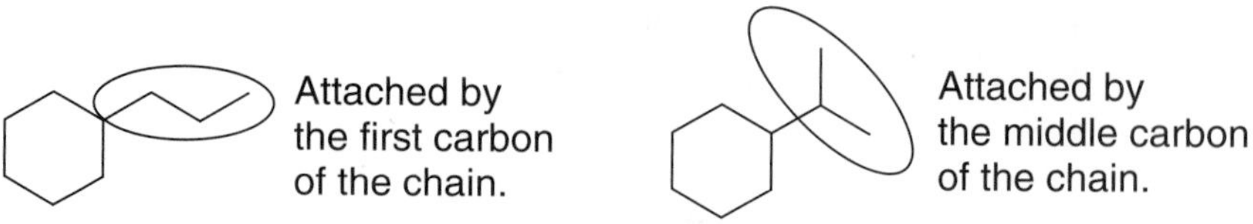

Attached by the first carbon of the chain.

Propyl

Attached by the middle carbon of the chain.

Isopropyl

It is still a chain of three carbon atoms, but it is attached to the parent chain differently than a propyl group is attached, so we call it isopropyl. This is an example of a branched substituent (branched, because it is not connected in one straight line to the parent chain, like a propyl group is).

Another important branched substituent to be familiar with is the *tert*-butyl group:

butyl   *tert*-butyl

The *tert*-butyl group is made up of four carbon atoms, just like a butyl group, but the *tert*-butyl group is not a straight line connected to the parent. Rather, the group has three methyl groups attached to one carbon, which is itself attached to the parent chain. So, we call this group *tert*-butyl.

There are two other ways to attach four carbon atoms to a parent chain (other than butyl and *tert*-butyl). As a small assignment, see if you can find their names in your textbook.

There is another important type of substituent that we need to cover. When we learned about functional groups, we saw that some compounds can have two functional groups. When this happens, we need to choose one of the functional groups to get the suffix, and the other functional group gets named as a substituent. To choose the functional group that gets the priority, go back to the section on functional groups and you will see the list of six functional groups (they are in order of priority, where a carboxylic acid always gets the priority). We need to know how to name any other functional groups in the molecule as substituents. The OH group is named –hydroxy- as a substituent. The NH_2 group is called –amino- if it is named as a substituent. A ketone is called –keto- as a substituent, and an aldehyde is called –aldo- as a substituent. Knowing how to name those four functional groups as substituents will probably cover you for anything you will see in your course.

Halogens are named as substituents in the following way: fluoro, chloro, bromo, and iodo. Essentially, we add the letter "o" at the end to say that they are substituents. If there are multiple substituents of the same kind (for example, if there are five chlorine atoms on the compound), we use the same prefixes that we used earlier when classifying the number of double and triple bonds:

di = 2   penta = 5

tri = 3   hexa = 6

tetra = 4

Each and every substituent needs to be numbered so that we know where it goes on the parent chain, but we will learn about this after we have finished going through the five parts of the name. At that time, we will also discuss in what order to place substituents in the name.

EXERCISE 5.28 In the following compound, identify all groups that would be considered substituents, and then indicate how you would name each substituent:

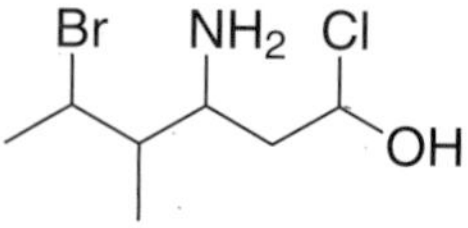

Answer First we must locate the functional group that gets the priority. Alcohols outrank amines, so the OH group is the priority functional group. Then, we need to locate the parent chain. There are no double or triple bonds, so we choose the longest chain containing the OH group:

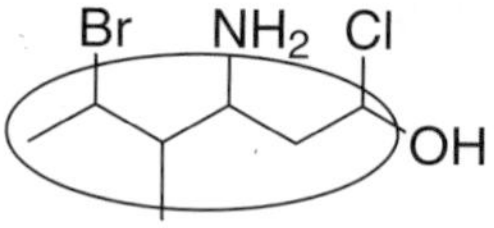

Now we know which groups must be substituents, and we name them accordingly:

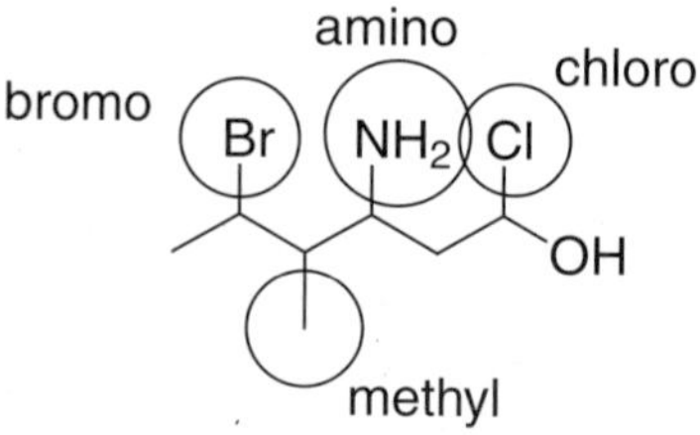

PROBLEMS For each of the following compounds, identify all groups that would be considered substituents, and then indicate how you would name each substituent.

5.29

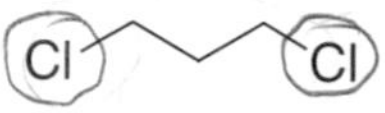

5.30

5.31

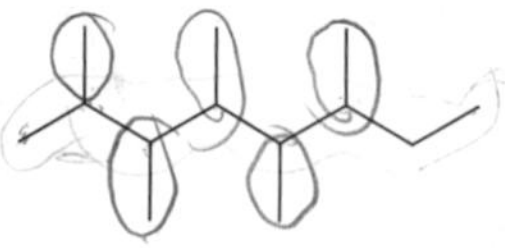

5.32

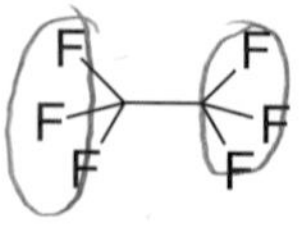

5.33

5.34

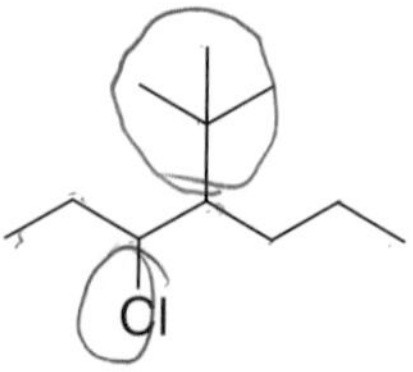

5.35

5.36

5.37

5.38

5.5 STEREOISOMERISM

Stereoisomerism	Substituents	Parent	Unsaturation	Functional group

Stereoisomerism is the first part of every name. It identifies the configuration of any double bonds or stereocenters. If there are no double bonds or stereocenters in the molecule, then you don't need to worry about this part of the name. If there are, you must learn how to identify the configuration of each. Identifying the configuration of a stereocenter requires a chapter to itself. You will need to learn what a stereocenter is, how to locate them in molecules, how to draw them, and how to assign a configuration (R or S). These topics will all be covered in detail in Chapter 7. At that time, we will revisit how to appropriately place the configuration in the name of the molecule. For now, you should know that configurations are placed here in the first part of a name.

Here we will focus on double bonds, which can often be arranged in two ways:

cis   trans

This is very different from the case with single bonds, which are freely rotating all of the time. But a double bond is the result of overlapping p orbitals, and double bonds *cannot* freely rotate (if you had trouble with this concept when you first learned it, you should review the bonding structure of a double bond in your textbook or notes). So there are two ways to arrange the atoms in space: cis and trans. If you compare which atoms are connected to each other in each of the two possibilities, you will notice that all of the atoms are connected in the same order. The difference is how they are connected *in 3D space*. This is why they are called stereoisomers (this type of isomerism stems from a difference of orientation in space—"stereo").

To name a double bond as being cis or trans, you need to have identical groups on *either side* of the double bond that can be compared to each other. If these identical groups are on the same side of the double bond, we call them cis. If they are on opposite sides, we call them trans:

| Two methyl groups are trans | Two fluorine atoms are cis | Two ethyl groups are trans | Two methyl groups are trans |

The two groups that we compare can even be hydrogen atoms. For example,

is trans because there are two hydrogen atoms not shown, and they are trans to each other:

But what do you do if you don't have two identical groups to compare. For example,

is not the same as

These compounds are clearly not the same. We cannot use cis/trans terminology to differentiate them, because we don't have two identical groups to compare. In situations when all four groups on the double bond are different, we have to use another method for naming them.

The other way of naming double bonds uses rules similar to those used in determining the configuration of a stereocenter (R versus S), so we will wait until the next chapter (when we learn about R and S), and then we will cover this alternative way of naming double bonds. The alternative method is far superior, because it can be used to name any double bond. In contrast, cis/trans nomenclature can be used only when we have two identical groups. The reason that we do not drop the cis/trans terminology altogether is probably based in deep-rooted tradition and usage of these terms.

There is one situation when we don't have to worry about cis/trans or E/Z because there aren't two ways to arrange the double bond. If we have two identical groups connected *to the same atom,* then we cannot have stereoisomers. For example,

is the same as

because there are two chlorine atoms connected to one carbon atom on one side of the double bond. Why are the two drawings the same? Remember that the carbon atoms of the double bond are sp^2 hybridized, and therefore trigonal *planar*. So, if we flip over the first drawing, we get the other drawing. They are the same thing. To see this, take two pieces of paper. Draw one of these compounds on one piece of paper, and draw the other compound on the second piece of paper. Then flip over one of the pieces of paper, and hold it up to the light so that you can see the drawing through the back side of the paper. Compare it to the other drawing and you will see that they are the same. If you try to do the same thing with some of the previous examples (that did have cis and trans stereoisomers), you will find that flipping the page over does not make the two drawings the same. This is a useful exercise, so take a few minutes and do it.

EXERCISE 5.39   Determine whether the double bond below is cis or trans:

Answer   Begin by circling the four groups attached to the double bond and try to name them:

You should always use this technique, because it will help you see when you have two groups that are the same. There are always four groups on the double bond (even if some of them are just hydrogen atoms). In this case, it helps us see that there are two isopropyl groups on the same side of the double bond. Therefore, the double bond is cis.

PROBLEMS   For each of the compounds below, determine whether the double bond is cis or trans.

5.40

5.41

5.42

5.43   D

5.44

5.45

5.6 NUMBERING

Stereoisomerism	Substituents	Parent	Unsaturation	Functional group

Numbering applies to all parts of the name

We're almost ready to start naming molecules. We finished learning about the individual parts of a name, and now we need to know how to identify where each of those pieces goes. For example, let's say you determine that the functional group is OH (therefore, the suffix is –ol), there is one double bond (-en-), the parent chain is six carbon atoms long (hex), there are four methyl groups attached to the parent chain (tetramethyl), and the double bond is cis. Now you know all of the pieces, but we must find a way to identify where all of the pieces are on the parent chain. Where are all of those methyl groups? (and so on). This is where the numbering system comes in. First we will learn how to number the parent chain, and then we will learn the rules of how to apply those numbers in each part of the name.

Once we have chosen the parent chain, there are only two ways to number it: right to left or left to right. But how do we choose? To number the parent chain properly, we begin with the same hierarchy that we used when choosing the parent in the first place:

Functional group
Double bond
Triple bond

If there is a functional group, then number the parent chain so that the functional group gets the lower number:

OH gets the number 2 instead of 5

If there is no functional group, then number the chain so that the double bond gets the lower number:

The double bond is 1 instead of 5

If there is no functional group or double bond, then number the chain so that the triple bond gets the lower number:

The triple bond is 1 instead of 5

If there is no functional group, double bond, or triple bond, then we should number the chain so that the substituent has the lower number:

Cl gets the number 3 instead of 4

If there is more than one substituent on the parent chain, then we should number the chain so that the substituents get the lowest numbers possible:

3,3,4-Trichloro instead of 3,4,4-trichloro

EXERCISE 5.47   For the compound below, choose the parent chain and then number it correctly:

Answer   To choose the parent chain, remember that we need to choose the longest chain containing the functional group:

To number it correctly, we need to go in the direction that gives the functional group the lowest number:

PROBLEMS   For each of the compounds below, choose the parent chain and number it correctly.

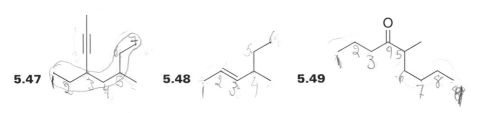

5.47   **5.48**   **5.49**

5.50 HO

5.51

5.52

5.53

5.54 OH

5.55

Now that we know how to number the parent chain, we need to see how to apply those numbers to the various parts of the name.

Functional Group The number generally gets placed directly in front of the suffix (for example, hexan-2-ol). If the functional group appears at the number 1, then the number does not need to be placed in the name (for example, hexanol). It is assumed that the absence of a number means that the functional group is at the number 1 position. When placing a number, it is OK to place the number at the beginning of the name if there are no other numbers in the name (from substituents or anything else). For example, 2-hexanol is the same as hexan-2-ol.

Unsaturation For double and triple bonds, the number indicates the lower number of the two carbon atoms. For example,

$$2 \quad 4 \quad 6$$
$$1 \quad 3 \quad 5$$

We use the number 2

The double bond is between C2 and C3, so we use C2 to number the double bond. So the example above is hex-2-ene (or 2-hexene, because there are no other numbers in the name, so it is OK to put the number in front). We treat triple bonds the same way.

If there are two double bonds in the molecule, then we must indicate both numbers; for example, hexa-2,4-diene, or 2,4-hexadiene. Every double and triple bond must be numbered.

Substituents The number of the substituent goes immediately in front of the substituent. Examples:

Cl

2-chlorohexane

3-methylpentane

This does not change if there are double bonds, triple bonds, or functional groups:

Cl OH

2-chlorohexan-2-ol

Cl OH

2-chlorohex-3-en-2-ol

If there are multiple substituents, then every substituent must be numbered:

2,3-**dichloro**hexane   2,2,4-**tri**methylpentane

If there are multiple substituents of different types, then we must alphabetize the substituents in the name. Consider the following example:

There are four types of substituents in the example above (chloro, fluoro, ethyl, and methyl). They must be alphabetized (c, e, f, m) (we do not count di, tri, tetra, etc. as part of the alphabetization system). So the compound above is called

2-chloro-3-ethyl-2,4-difluoro-4-methylnonane

Note that two numbers are always separated by commas (2,4 in example above) but letters and numbers are separated by dashes (2-chloro-3-ethyl . . .).

Stereoisomerism   If there are any double bonds, we place the term cis or trans at the beginning of the name. If there is more than one double bond, then we need to indicate cis or trans for each double bond, and we must number accordingly (for example, 2-cis-4-trans . . .). If there are any stereocenters, here is where we would indicate them; for example, (2R,4S). Stereocenters are placed in parentheses. We will see more of this when we learn about stereocenters in the upcoming chapters.

There it is. A lot of rules. No one ever said nomenclature would take 10 minutes to learn, but with enough practice, you should get the hang of it. Let's now take everything we have learned and apply it to solving some problems:

EXERCISE 5.56   Name the following compound:

Answer   We go through the five parts of the name backward. So we start by looking for the functional group. We see that there is a ketone. So we know the end of the name will be –one.

Next, we look for unsaturation. There is a double bond here, so there will be –en- in the name.

Next we need to name the parent. We locate the longest chain that includes the functional group and double bond. In this example, it is an obvious choice. The parent has 7 carbon atoms, so the parent is –hept-.

Next we look for substituents. There are two methyl groups and two chlorines. We need to alphabetize, and c comes before m, so it will be dichlorodimethyl.

Next we look for stereoisomerism. The double bond in this molecule has two chlorine atoms on opposite sides, so it is trans. This part of the name (*trans*) is generally italicized. So far, we have

   trans-dichlorodimethylheptenone

Now that we have figured out all of the pieces, we must number everything. We need to number the parent to give the functional group the lowest number, so the numbering, in this example, will go from left to right. This puts the functional group at the number 2 position, the double bond at the 4 position, the chlorines at 4 and 5, and the methyls at 6 (both of them). So the name is

   trans-4,5-dichloro-6,6-dimethylhept-4-en-2-one

PROBLEMS   Name each of the following compounds. (Ignore stereocenters for now. We will focus on stereocenters in the upcoming chapters.)

5.57   Name: ~~trans-5-ethyl~~ – 4-methyloct-2-ene

5.58   Name: 4-ethylnonan-3-ol

5.59   Name: 4,4-dimethylhex-2-yne

5.60   Name: 4,4-dimethylcyclohexanone

5.61   Name: 2-chloro-4-fluoro-3,3-dimethyl hexane

5.62   Name: cis-3-methylhex-2-ene

5.63   Name: 2-ethyl pentamine

5.64   Name: 2-propylpentanoic acid   hanol

5.65   Name: trans-oct-2-en-4-ol

5.66   Name: trans-5-chloro-6-fluoro-2-en 5-6 dimethyloct-2-ene

5.7 COMMON NAMES

In addition to the rules for naming compounds, there are also some common names for some simple and common organic compounds. You should be aware of these names to the extent that your course demands this of you. Each course will be different in terms of how many of these common names you should be familiar with. Here are some examples:

IUPAC name: methanoic acid
Common name: formic acid

IUPAC name: ethanoic acid
Common name: acetic acid

IUPAC name: methanal
Common name: formaldehyde

IUPAC name: ethanal
Common name: acetaldehyde

IUPAC name: ethene
Common name: ethylene

IUPAC name: ethyne
Common name: acetylene

Most of these examples are so common, that it is quite rare to hear someone refer to these compounds by their IUPAC names. Their common names are much more "common," which is why we call them common names.

Ethers are typically called by their common names. The group on either side of the oxygen is named as a substituent before the term ether. Examples:

Diethyl ether
(also just ethyl ether)

Dimethyl ether

The IUPAC method would be to treat the oxygen like a carbon and then indicate where the oxygen is with the term oxa-. So diethyl ether would be 3-oxapentane, but no one calls it that. Everyone calls it diethyl ether (or just ethyl ether). It is not a bad idea to familiarize yourself with all of the common names listed in whatever textbook you are using.

5.8 GOING FROM A NAME TO A STRUCTURE

Once you have completed all of the problems in this chapter, you will find that it is much easier to draw a compound when you are given the name than it is to name a compound that is drawn in front of you. It is easier for the following reason: when naming a compound, there are a lot of decisions you need to make (which functional group has priority, what is the parent chain, how the chain should be numbered, in what order to put the substituents in the name, etc.). But when you have a name in front of you and you need to draw a structure, you do not need to make any of these decisions. Just draw the parent and then start adding everything else to it according to the numbering system provided in the name.

For practice, make a list of the answers to problems 5.58–5.67. This list should just be names. Wait a few days until you cannot remember what the structures looked like and then try to draw them based on the names. You can also use your textbook for more examples.

From this point on, I will assume that I can say words like 2-hexanol and you will know what I mean. That is what your textbook will do as well, so now is the time to master nomenclature.

CHAPTER 6

CONFORMATIONS

Molecules are not inanimate objects. Unlike rocks, they can twist and bend into all kinds of shapes, very much like people do. We have limbs and joints (elbows, knees, etc.) that give us flexibility. Although our bones might be very rigid, nevertheless, we can achieve a great range of movement by twisting our joints in different ways. Molecules behave the same way. Once you learn about the general types of joints that molecules have and the ranges of motion available to those joints, you will then be able to predict how individual molecules can twist around in space. Why is this important?

Let's stay with the people analogy. Think about how many common positions you assume every day. You can sit, you can stand, you can lean against something, you can lie down, you can even stand on your head, and so on. Some of these positions are comfortable (such as lying down), while other positions are very uncomfortable (such as standing on your head). There are some activities, like drinking a glass of water, which can be done only in certain positions. You cannot drink a glass of water while lying down or standing on your head.

Molecules are very similar. They have many positions into which they can twist and bend. There are some comfortable positions (low in energy), while other positions are uncomfortable (high in energy). The various positions available to a molecule are called *conformations*. It is important to be able to predict the conformations available to molecules, because there are certain activities that can be performed only in specific conformations. Just as a person standing on her head cannot drink a glass of water, so, too, molecules cannot undergo some reactions from certain conformations. Just as you can run only when you are in a standing position, molecules can often undergo certain reactions from only one specific conformation. If the molecule cannot twist into the conformation necessary for a reaction to take place, then the reaction will not happen. So you can understand why you will need to be able to predict what conformations are available to molecules—so that you can predict when reactions can and cannot happen.

There are two very important drawing styles that show conformations and give us the power to predict what conformations are available to different types of molecules: Newman projections and chair conformations:

Newman projection   Chair

We begin this chapter with Newman projections.

6.1 HOW TO DRAW A
NEWMAN PROJECTION

Before we can talk about drawing Newman projections, we need to first review one aspect of drawing bond-line structures that we did not cover in Chapter 1. To show how groups are positioned in 3D space, we often use wedges and dashes:

In the bond-line drawing above, the fluorine atom is on a wedge and the chlorine atom is on a dash. The wedge means that the fluorine is coming out toward us in 3D space, and the dash means that the chlorine is going away from us in 3D space. Imagine that all four carbon atoms in the molecule above are positioned in the plane of the page. If you look at this page from the left side (so the page looks one-dimensional), you would see the fluorine sticking out of the page to your right and the chlorine sticking out of the page to the left.

Do not be confused by whether the dash is drawn on the right or left:

is the same as

In both drawings above, the chlorine is on a dash and the fluorine is on a wedge, so these drawings are the same. In reality, both the chlorine and the fluorine should be drawn straight up—the chlorine goes straight up and behind the page, while the fluorine goes straight up and in front of the page. If we drew it like that, we would not be able to see the chlorine because the fluorine would be blocking it (the way the moon blocks the sun in a solar eclipse). To clearly see both groups, we move one of the groups slightly to the left and the other slightly to the right. It does not matter which is on the right and which is on the left. All that matters is which one is on the wedge and which one is on the dash.

Now that we understand what the dash and wedge mean, let's consider what the molecule would look like from a slightly different angle:

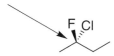

Imagine looking at the molecule from the angle shown by the arrow above. If you are not sure what angle we are talking about, try doing the following: Turn your book so that it is facing your stomach instead of your head. Now turn the page so you are looking at it from the side, like we did before. You should be looking down the path of this arrow at the molecule. In this view, you are looking directly down a carbon–carbon bond, where one carbon is in front and one is in back:

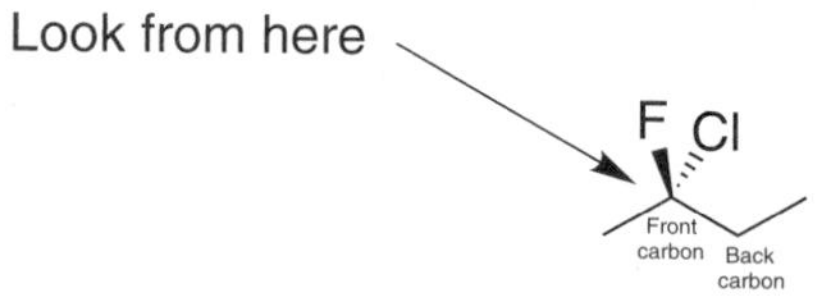

In this view, you will see three groups connected to the front carbon atom. You should expect to see a fluorine atom sticking out of the page to the right and a chlorine atom sticking out of the page to the left. You would also see a methyl group pointing straight down. This is what it would look like from that view:

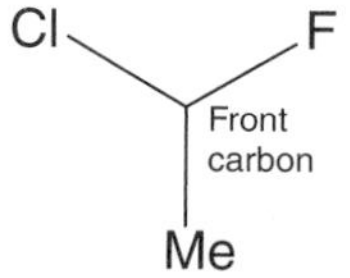

You would see the three groups like this, and you would not see the back carbon, because the carbon in front would be covering it up (again, like the moon covers the sun during a solar eclipse). Let's try to draw that back carbon that we cannot see, and by convention, we will draw it as a big circle:

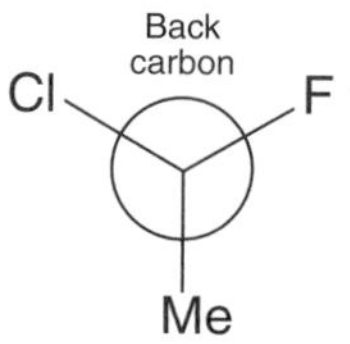

Now we place the three groups that are on the back carbon into our drawing. There is one methyl group and two hydrogen atoms. If we put them into our drawing, then we get our Newman projection. It looks like this:

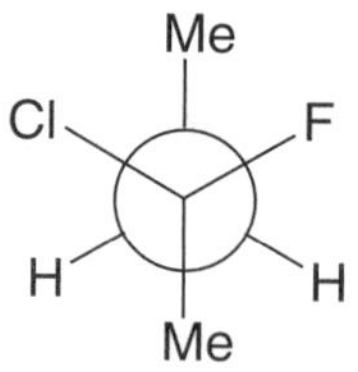

It is important that you can see what this drawing represents, because we cannot move forward until you see it very clearly. We are essentially looking at a carbon–carbon bond, focusing on the three groups attached to each carbon atom. The central point in our Newman projection (where the lines to the Cl, F, and Me meet each other) is the first carbon. The big circle in the back is our back carbon. All at once you can see all six groups (the three connected to the front carbon and the three connected to the back carbon). So a Newman projection is another way of drawing the compound we showed earlier:

Let's use one more analogy to help us understand this. Imagine that you are looking at a fan that has three blades. Behind this fan, there is another fan that also has three blades. So you see a total of six blades. If both fans were spinning, and you started taking photographs, you might find some photos where you can clearly see all six blades, and other photos where the three blades in the front are blocking our view of the three blades in the back. In this last case, the three blades in front would be *eclipsing* the three blades in back.

This is where the analogy helps us understand why Newman projections are useful. The bond connecting the two carbon atoms is a single bond that can freely rotate. Sometimes, you can see all six groups because they are *staggered*. But other times, you can't see the groups in the back because they are being *eclipsed* by the three groups in the front:

Think of the front carbon and its three groups as one fan, and the back carbon and its three groups as a different fan. These two fans can spin independently of each other, which gives rise to many different possible conformations. This is why Newman projections are so incredibly powerful at showing conformations. They are drawn in a way that is perfect for showing the various conformations that arise as an individual single bond rotates.

It gets a little more complicated when we realize that every single bond in every molecule can freely rotate, giving rise to a very large number of conformations for molecules. We can avoid that kind of complexity by focusing just on one particular bond, and the various conformations that arise from free rotation of just that bond. If we can learn how to do that, then we can do that for the part of the molecule that is undergoing a reaction and we do not need to concern ourselves with the rest of the molecule.

EXERCISE 6.1 Draw a Newman projection of the following compound looking from the perspective of the arrow:

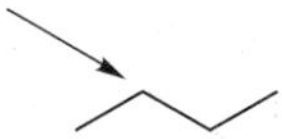

Answer The first thing to realize is that there are groups on dashes and wedges that have not been shown. They are the hydrogen atoms. We did not focus on drawing the dashes and wedges in Chapter 1, but the hydrogen atoms are, in fact, on dashes and wedges. We learned in the chapter on geometry that these carbon atoms would be classified as sp^3 hybridized, and, therefore, their geometry is tetrahedral. That means the hydrogen atoms are coming in and out of the page:

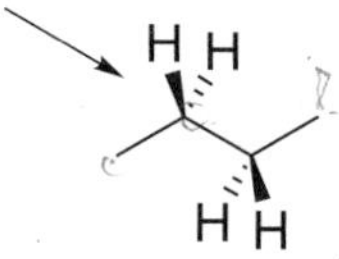

Now we draw the front carbon with its three groups. Looking along the direction of the arrow, we see a hydrogen up and to the left, another hydrogen up and to the right, and then a methyl group pointing straight down. So we draw it like this:

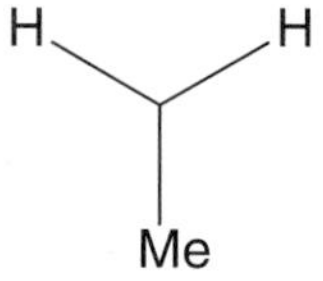

Next we draw in the back carbon as a large circle and we look at all three groups attached to it. There is a methyl group pointing straight up and then two hydrogen atoms pointing left and right. So the answer is

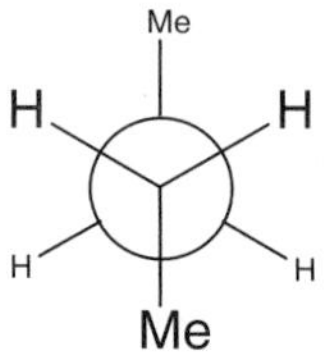

PROBLEMS Draw Newman projections for each of the following compounds. In each case the skeleton of the Newman projection is drawn for you. You just need to fill in the six groups in their proper places.

6.2

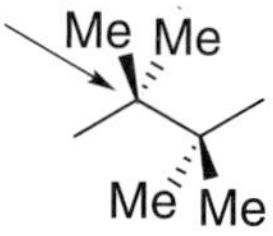
Me Me
Me Me

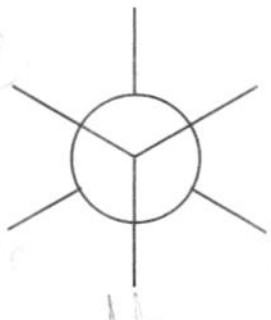

Answer

6.3

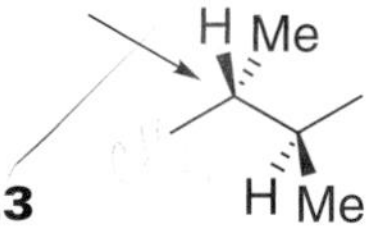
H Me
H Me

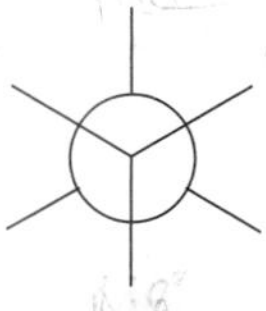

Answer

6.4

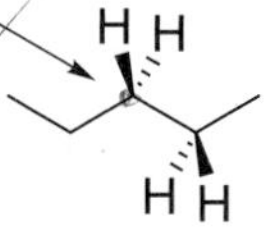
H H
H H

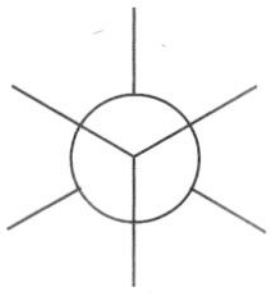

Answer

6.5

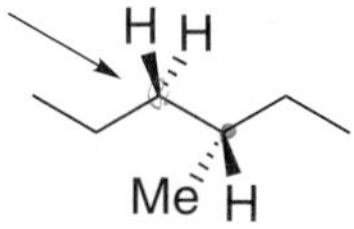
H H
Me H

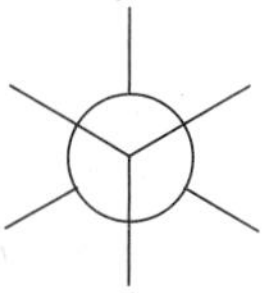

Answer

6.6

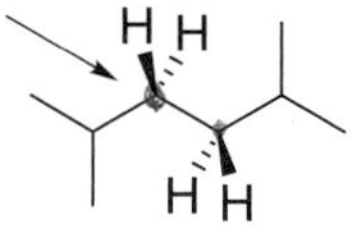
H H
H H

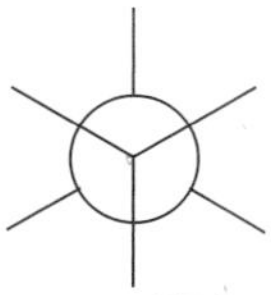

Answer

6.7

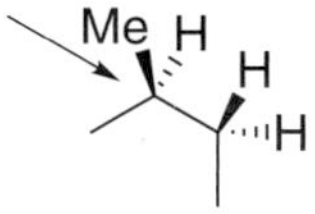
Me H
H
H

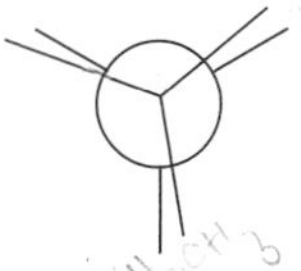

Answer

6.2 RANKING THE STABILITY OF NEWMAN PROJECTIONS

We have seen that Newman projections are a powerful way to show the different conformations of a molecule. We mentioned earlier that there are staggered conformations and eclipsed conformations. In fact, there are three staggered and three eclipsed conformations. Let's draw all three staggered conformations of butane. The best way to do this is to keep the back carbon motionless (so the fan in the back is not spinning), and let's slowly turn the groups in the front (only the front fan is spinning):

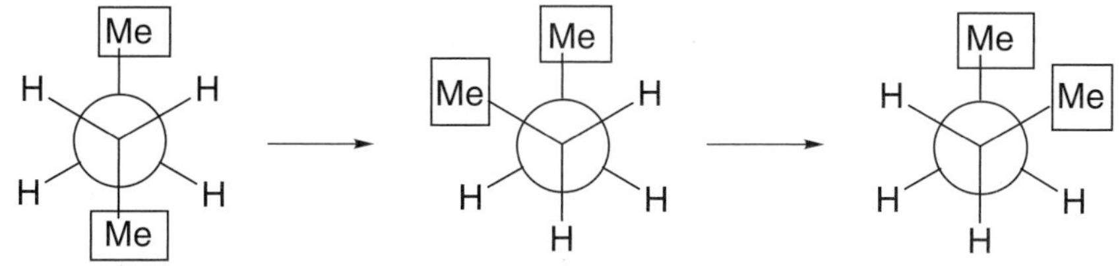
Me
H
H
H
H
Me
Me
Me
H
H
H
H
Me
H
Me
H
H
H

Anti Gauche Gauche

Look at the first drawing above and notice the placement of the methyl group at the bottom. If we rotate the front carbon clockwise with all three of its groups, this methyl ends up in the top left (as seen in the second drawing). Then we rotate one more time to get the third drawing. Rotating one more time regenerates the first drawing. In the first drawing, the methyl groups are as far away from each other as possible, which is the most stable conformation, called the *anti* conformation. The other two drawings both have the methyl groups near each other in space. They feel each other and they are a bit crowded. This interaction is called a *gauche* interaction, which makes these conformations a little bit less stable than the anti conformation.

If we were to go back to our comparison between molecules and people, we would say that the anti conformation would be like lying down in a bed, and the gauche conformations are both like sitting in a chair. All of these are comfortable positions, but lying down is the most comfortable. The anti conformation is the most stable.

Now let's look at the three eclipsed conformations of butane. Again, let's keep the back where it is, and let's just rotate the front carbon with its three groups:

These conformations are all high in energy relative to the staggered conformations. All of the groups are eclipsing each other, so they are very crowded. All three of these would be like standing on your head, which is extremely uncomfortable. But the middle one is the most unstable, because the two methyl groups (the two largest groups) are eclipsing each other. This would be like standing on your head without using your hands for help—now that *really* hurts. So, all of these are high in energy, but the middle one is the most unstable.

To summarize, the most stable conformation will be the staggered conformation where the large groups are as far apart as possible (anti), and the least stable conformation will be the eclipsed conformation where the large groups are eclipsing each other.

EXERCISE 6.8   Draw the most stable and the least stable conformations of the following compound, using Newman projections looking down the following bond:

Answer   We begin by drawing the Newman projection that we would see when we look down the bond indicated. This Newman projection will have a methyl group and two hydrogen atoms connected to the front carbon, and there will be an ethyl group and two hydrogen atoms connected to the back carbon, in the following way:

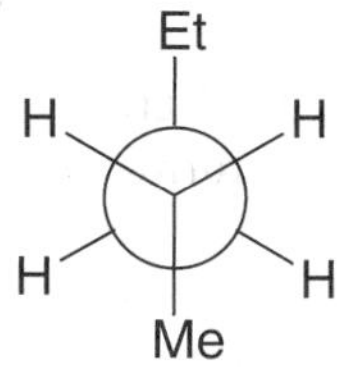

Now we decide how to rotate the front carbon so as to provide the most stable conformation. The most stable conformation will be a staggered conformation with the two largest groups anti to each other, so in this case, we do not need to rotate at all. The drawing that we just drew is already the most stable conformation, because the methyl and ethyl groups are anti to each other:

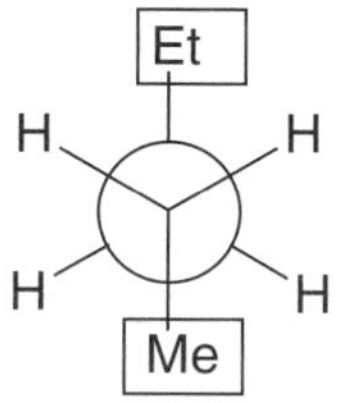

To find the least stable conformation, we need to rotate the front carbon atom and consider all three eclipsed conformations. The least stable conformation will be the one with the two largest groups eclipsing each other:

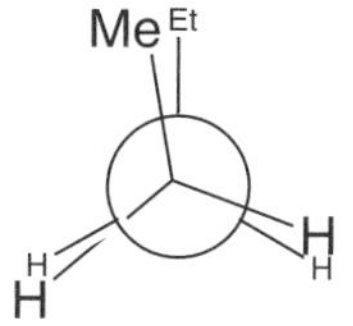

PROBLEMS Draw the most stable and the least stable conformations for each of the following compounds. In each case, fill in the groups on the Newman projections below.

6.9

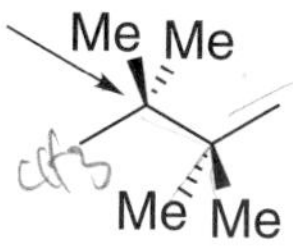

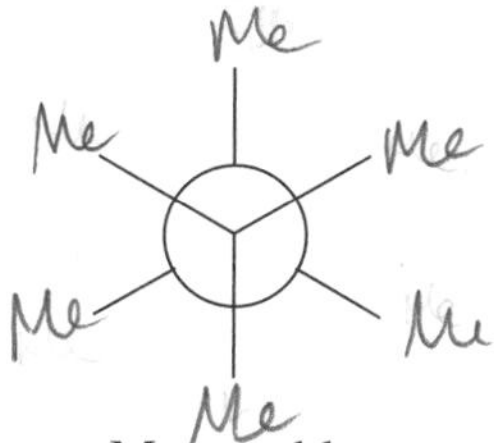

Most stable 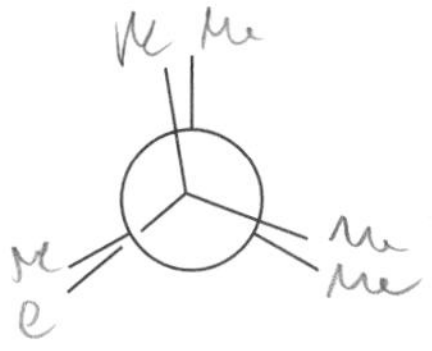   Least stable

6.10

H Me
H Me

Most stable   Least stable

6.11

H H
H H

Most stable   Least stable

6.12

H H
Me H

Most stable   Least stable

6.13

H H
H H

Most stable   Least stable

6.14

Me H
H
H

Most stable   Least stable

6.3 DRAWING CHAIR CONFORMATIONS

An interesting case of conformational analysis comes to play when we consider a six-membered ring (cyclohexane). There are many conformations that this compound can adopt. You will see them all in your textbook: the chair, the boat, the twist-boat. The most stable conformation of cyclohexane is the chair. We call it a chair, because when you draw it, it looks like a chair:

You can almost imagine someone sitting on this structure, as if it were a beach chair. Most students have a difficult time drawing the chair and its substituents correctly. In this section, we will focus on learning how to draw the chair correctly. It is very important, because we cannot move on to see more about chairs until we know how to draw them.

You will need to practice this, step by step. Begin by drawing a *very wide* V:

Next, draw a line from the top right of the V, going down at a 60-degree angle, and stop a little bit to the right of an imaginary line coming straight down from the center of the V:

Next, draw a line parallel to the left side of the V and stop a little bit to the right of an imaginary line coming straight dawn from the top left of the V:

Then, start at the top left of the V and draw a line parallel to the line all the way on the right side, going down exactly as low as that line goes:

Finally, connect the last two lines:

Please don't *ever* draw a chair like this:

When you draw the chair sloppily (as so many students do), it makes it impossible to place the substituents correctly on the ring. And that's when you start losing silly points on your exam. So, take the time to practice and learn how to draw it properly. Practice in the space below:

Now we can start drawing in the substituents. Start with the top right corner and draw a line going straight up:

Then go around the ring and draw a straight line at each carbon atom, alternating be-tween up and down:

These six substituents are called *axial* substituents. They go straight up and down, in the order shown above.

Next we need to learn how to draw the *equatorial* substituents. These are the substituents pointing out toward the sides. There are also six of them. Each one is drawn so that it is parallel to the two bonds from which it is once removed:

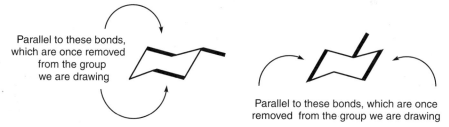

Parallel to these bonds, which are once removed from the group we are drawing

Parallel to these bonds, which are once removed from the group we are drawing

We go all the way around the ring like this, until we have drawn all of the equatorial groups:

Now we know how to draw all twelve substituents, but remember that if we draw a line and don't put anything at the end of that line, then this implies a methyl group. So, unless we are referring to dodecamethylcyclohexane (that's 12 methyl groups), we should really draw hydrogen atoms at the end of these lines:

Generally, we do not have to draw all 12 lines and place hydrogen atoms there. Remember how bond-line drawings work: if we don't draw any groups at all, it is assumed that there are hydrogen atoms. We are going through this exercise because it is important to know *how* to draw all 12 substituents. We will see in the next section that in most problems you will need to draw only a few of them. Which ones you draw will depend on the problem. So the only way to be sure that you can draw whatever the problem throws at you is to master drawing *all* of them. *Never* draw groups like this:

Very bad

Drawings like this will cause you to lose serious points on exams (not to mention the fact that a drawing like this defeats the whole purpose of a chair drawing—the exact positioning of the substituents is very important).

Use the space below to practice drawing a chair with all twelve substituents. When you are finished, label every substituent as axial or equatorial. Do not move on to the next section until you can do this.

6.4 PLACING GROUPS ON THE CHAIR

Now we need to see how to draw a chair with proper placement of the groups when we are given a regular hexagon-style drawing:

is the same as

Before we can get started, we need to remember what the dashes and wedges mean in the hexagon-style drawing. Remember that wedges are coming out toward you, and dashes are going back away from you. So, at each of the six carbon atoms of the ring, there are two groups—one coming out at you and one going away from you. If the groups are not drawn, then it means that there are two hydrogen atoms, one coming out and one going away.

Now let's introduce some new terminology. This is not scientific terminology, and you won't find it in any textbook, but it will help you master the task at hand. Anything coming off of the ring that is on a wedge, we will call *up,* because the group is coming up above the ring. Any group on a dash, we will call *down,* because the group is going down below the ring. So in the example above, Br is up and Cl is down.

Now let's apply the same terminology to the groups on a chair. Each carbon atom has two groups, one pointing above the ring (up) and one pointing below the ring (down):

You can do this for every carbon (and you should try on the drawing above), and you will see that each carbon has two groups (up and down). It is important to realize that there is *no correlation* between up/down and axial/equatorial. Look at the drawing above. For one of the carbon atoms, the up position is axial. For the other carbon atom showing its groups, the up position is equatorial. Take a close look at the two equatorial positions shown above. One of them is up and the other is down.

Now we are ready to draw a chair when we are given a hexagon. Let's work through the example that we started with:

We begin by placing numbers. These are not the same as the numbers that we used in naming compounds. These are just numbers that help us draw the chair with the groups in the right place. It does not matter where we start or which direction we go in, so let's just say that we will always start at the top and go clockwise:

Now, we draw a chair and we put numbers on the chair also. We can start our numbers anywhere on the chair, but we *must* go in the same direction that we did in the hexagon. If we went clockwise there, then we must go clockwise here. To avoid a mistake, let's just say that we will always go clockwise from now on:

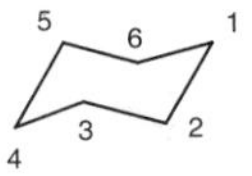

Now we know where to put in the groups. Br is on the carbon numbered 1, and Cl is on the carbon numbered 3. This brings us to the up/down system. Draw the chair, showing both positions (up and down) at each of the carbon atoms where we need to place a group:

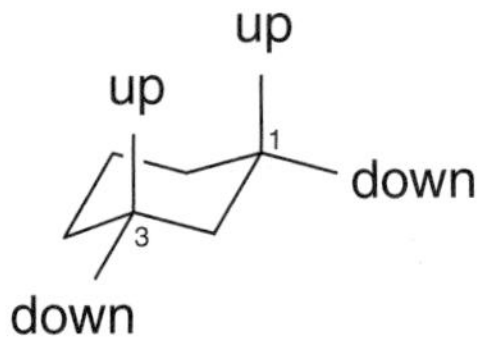

Look at the hexagon drawing again and ask whether each group is up or down. Br is on a wedge, so it is up. Cl is on a dash, so it is down. Now we are ready to put the groups into our chair drawing:

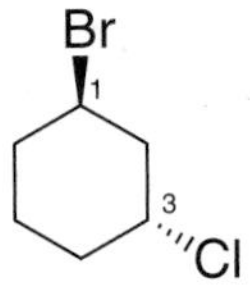

is the same as

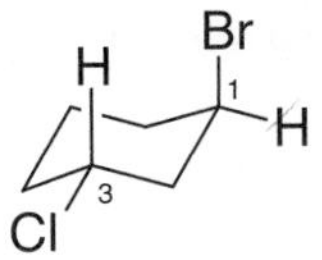

That's all there is to it. To review, we need to draw the chair, number the chair and hexagon (both clockwise), determine where the groups go, determine whether they are up or down, and then draw them in. Let's get some practice now.

EXERCISE 6.15 Draw a chair conformation of the compound below:

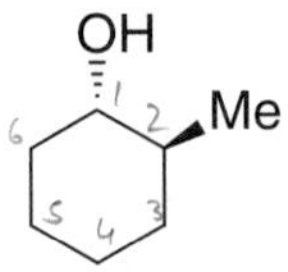

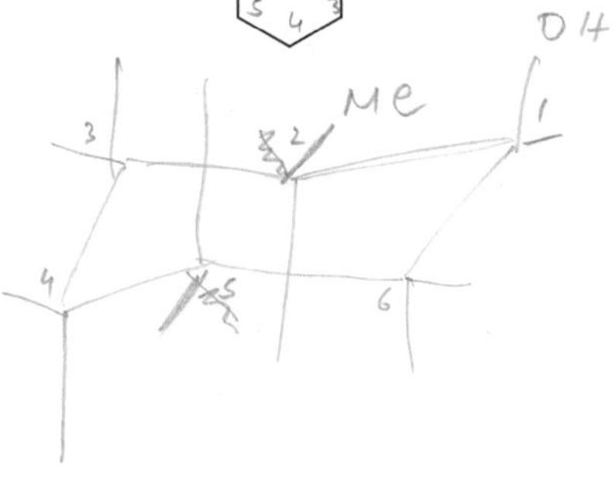

Answer   Begin by numbering the hexagon at the first group and then going clockwise. This puts the OH at the position numbered 1 and the Me at the position numbered 2:

OH
Me

Next draw the chair, number it going clockwise, and then put in the up and down positions at the carbons numbered 1 and 2:

up
down
up
down

Finally, place the groups where they belong. The OH should be at the number 1 position in the down position (because it was on a dash in the hexagon drawing), and the Me should be at the number 2 position in the up position (because it was on a wedge):

H
OH
Me
H

The example above illustrates an important point. Take a close look at the Me and the OH in the hexagon drawing. One is on a wedge and one is on a dash. We call this relationship *trans* (when you have two groups that are on opposite sides of the ring). If the two groups had been on the same side (either both on wedges, or both on dashes), then we would have called them *cis*. So in the example above, the groups are trans to each other. Now take a close look at the chair drawing we just drew above. The OH and Me don't look trans in this drawing. In fact, they look like they are cis, *but they are trans.* The OH is in the down position and the Me is in the up position.

It will become very clear that these groups are trans to each other when we learn how to draw the chair after it has "flipped" to give us a new chair drawing. We will see this in the next section. For now, let's get some practice drawing the first chair correctly.

PROBLEMS   Draw a chair conformation for each of the compounds below.

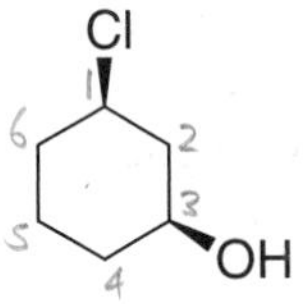

6.16

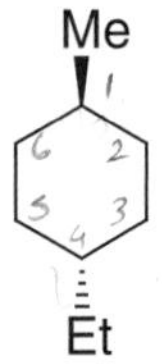

6.17

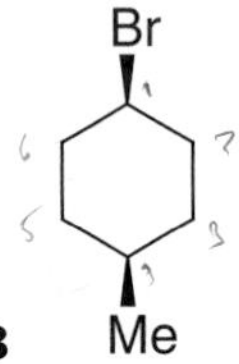

6.18

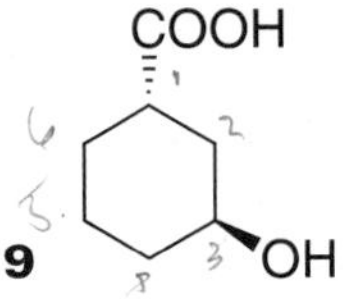

6.19

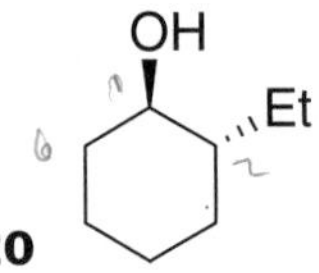

6.20

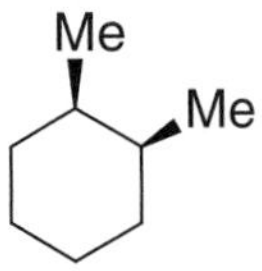

6.21

6.5   RING FLIPPING

Ring flipping is one of the most important aspects of understanding chair conformations, yet students commonly misunderstand this. Let's try to avoid the mistake
by starting off with what ring flipping is not. It is *not* simply turning the ring over:

It makes sense why students think that this would be a flip—after all, this is the
common meaning of the word "flipping." But we are talking about something very
different when we say that rings can flip. Here is what we really mean:

Notice that in the drawing on the left, the left side of the chair is pointing down. In
the drawing on the right, the left side of the chair is pointing up. This is a different
chair. Also notice that the chlorine went from being in an equatorial position to being
in an axial position. This is a critical feature of ring flips. When performing a ring
flip, all axial positions become equatorial, and all equatorial positions become axial.

Let's consider an analogy to help us get through this. Imagine that you are
walking down a long hallway. Your hands are swaying back and forth as you walk,
as most people do with their hands when they walk. One second, your left hand is in
front of you and your right hand is behind you; the next second, it switches. Your
hands switch back and forth with every step you take. The cyclohexane ring is doing
something similar. It is moving around all of the time, flipping back and forth between two different chair conformations. So all of the substituents are constantly
flipping back and forth between being axial and being equatorial.

There is one more important feature to recognize. Let's go back to the example above with the chlorine. We said that the chair flip moves the chlorine from an
equatorial position into an axial position. But what about the up/down terminology?
Let's see:

Notice that the chlorine is down all of the time. In other words, up/down is *not* some-thing that changes during a ring flip, but axial/equatorial does change during a ring flip. This proves that there is no relationship between up/down and axial/equatorial. If a substituent is up, then it will stay up all of the time, throughout the ring flipping process.

So now we can understand that a common hexagon-style drawing represents a molecule that is flipping back and forth between two chair conformations. The hexa-gon drawing shows us which substituents are up and which are down. That never changes. But whether those groups are axial or equatorial will depend on which chair you are drawing. So far, we have learned how to draw only one of these chairs. Now we will learn how to draw the other.

The process for drawing the skeleton of the chair is very similar to how we did it before. The only difference is that we draw our lines in the other direction. When we drew our first chair, we followed these steps:

Step 1   Step 2   Step 3   Step 4   Step 5

Now, to draw the other chair, we follow these steps:

Step 1   Step 2   Step 3   Step 4   Step 5

Compare the method for drawing the second chair to the method for drawing the first. The key is in step 2. If you compare step 2 for the first and second chair, every-thing else should flow from there. Use the following space to practice drawing the second chair:

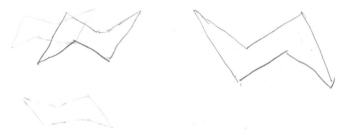

Now, let's make sure you know how to draw the substituents. The rules are the same as before. All axial positions are drawn straight up and down, alternating:

and all equatorial positions are drawn parallel to the two bonds that are once removed:

Parallel to these bonds, which are once removed
from the group we are drawing

PROBLEM

6.22 In the space below, practice drawing the second chair, showing all 12 substituents.

Let's now go back and review, because it is important that you understand the following points. When we are given a hexagon-style drawing, the drawing shows us which positions are up and which positions are down. No matter which chair we draw, up will always be up, and down will always be down. There are two chair conformations for this compound, and the molecule is flipping back and forth between these two conformations. With each flip, axial positions become equatorial positions and vice versa. Let's see an example.

Consider the following compound:

Notice that there are two groups on this ring. Cl is down (because it is on a dash), and Br is up (because it is on a wedge). There are two chair conformations that we can draw for this compound:

In both chair conformations, Cl is down and Br is up. The difference between these drawings is the axial/equatorial positions. In the conformation on the left, both groups are equatorial. In the conformation on the right, both groups are axial.

So any hexagon-style drawing will have two chair conformations. Now let's focus on making sure you can draw both conformations for any compound. We already saw in the last section how to draw the first one. We used a numbering system to determine where to put the groups, and we used the up/down terminology to figure exactly how to draw them (whether to draw them as equatorial or axial). To draw the second chair, we simply follow the same procedure. We begin by drawing the skeleton for the second chair (this is where you begin to see the difference between the chairs):

Skeleton for
first chair

Skeleton for
second chair

Once we have drawn the skeleton, we number the carbons going clockwise. Then we place the groups in the correct positions, making sure to draw them in the correct direction (up or down). So we can really use this method to draw both chairs at the same time. Let's do an example.

EXERCISE 6.23 Draw both chair conformations for the following compound:

OH
Me

Answer Begin by numbering the hexagon at the first group and then going clockwise. This puts the OH at the position numbered 1 and the Me at the position numbered 2.

OH
6 1 2 Me
5 3
4

Next draw both chair skeletons and number them going clockwise. Then put in the up and down positions at the carbons numbered 1 and 2:

up
5 6 1
down
4 3 2 up
down

up
4 5 6
up
3 2 1
down down

Finally, place the groups where they belong in both chairs. The OH should be at the number 1 position in the down position (because it was on a dash in the hexagon drawing), and the Me should be at the number 2 position in the up position (because it was on a wedge):

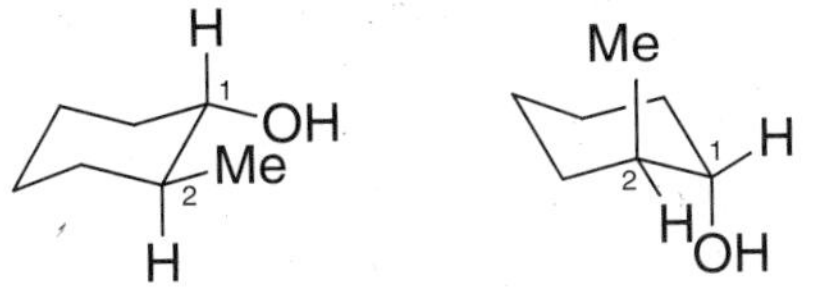

When we redraw these compounds without showing any numbers or hydrogen atoms, it is clear to see that we need to go through these steps methodically because the relationship between these two conformations is not so obvious:

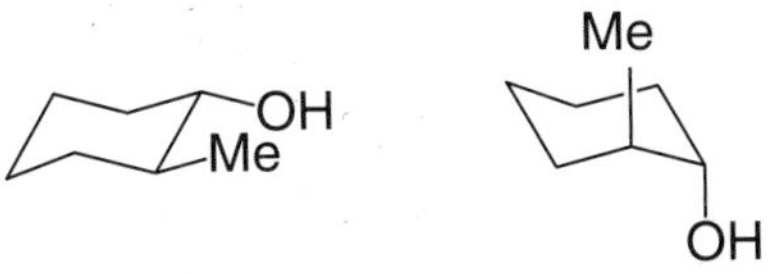

PROBLEMS For each of the compounds below, draw both chair conformations.

6.24

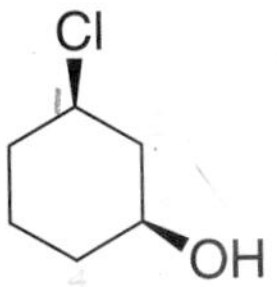

6.25 Et

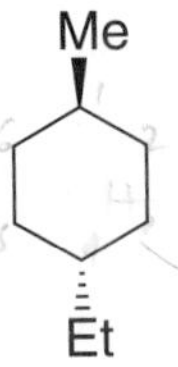

6.26 Me

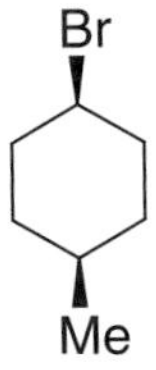

6.27

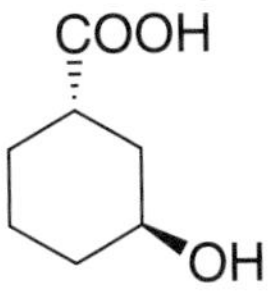

6.28

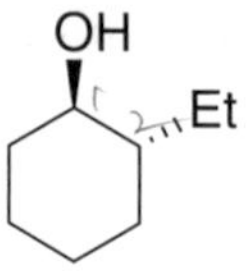

6.29

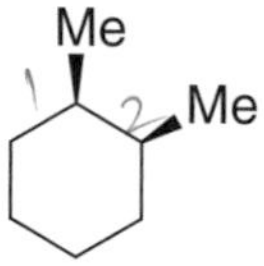

Sometimes, we might be given one chair conformation and be asked to draw the second chair conformation. Again, we use numbers to help us out. Let's see an example:

EXERCISE 6.30 Below you will see one chair conformation of a substituted cyclohexane. Draw the other chair (i.e., do a ring flip):

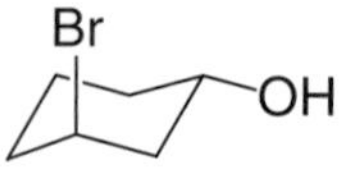

Answer   Begin by numbering the first chair. Start on the right side of the chair, and put a 1 at the first group. Then go clockwise. This puts the Br at the position numbered 3.

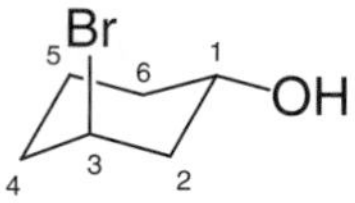

Notice that the OH is down and the Br is up.

Next draw the skeleton for the second chair. Begin numbering on the right side again, making sure to go clockwise. Then draw the down position at the 1 position, and draw the up position at the 3 position:

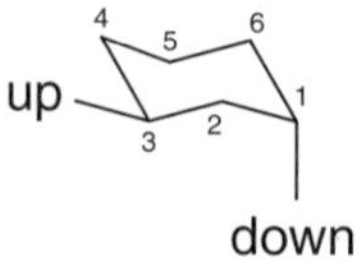

Finally, place the groups where they belong:

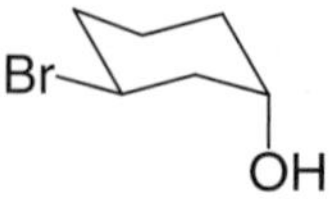

PROBLEMS For each chair conformation shown below, do a chair flip and draw the other chair conformation.

6.31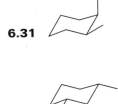

6.32

6.33

6.34

6.35

6.36

6.6 COMPARING THE STABILITY OF CHAIRS

Once you have drawn both chair conformations for a substituted cyclohexane, you should be able to predict which conformation is more stable. This is where it gets important for reactivity. Imagine that you are learning about a reaction that can proceed only if a certain group is in an axial position (you will learn about a reaction like this very soon—it is called E2). You already know that the groups are flipping back and forth between axial and equatorial positions (as it goes back and forth from one chair to the other). But what if one of the chairs is so unstable that the ring is spending 99% of its time in the other chair conformation? Then the question becomes, where is the important group in the stable chair conformation? Is it axial or equatorial? If it is equatorial, then the reaction can't happen. It could happen only during the 1% of the time that the group is in the axial position, so the reaction would be very slow. However, if the group is in an axial position 99% of the time, then the reaction will happen very quickly.

So you can see that it is important to understand what makes chair conformations unstable. There is really only one rule you need to worry about: a chair will be more stable with a group in an equatorial position, because it is not bumping into anything (this bumping is called *steric hinderance*). Axial positions are bumping into other axial positions, but equatorial positions are not:

The larger the group, the more it will prefer to be equatorial. So, a *tert*-butyl group will spend almost all of its time in an equatorial position. This essentially "locks" the ring in one conformation and prevents the ring from flipping (the truth is that the ring is still flipping, but the ring is spending more than 99% of its time in the more stable chair conformation):

So what happens if you also have a chlorine atom on the ring that is axial while the *tert*-butyl group is equatorial?

This will essentially lock the ring in the chair conformation that puts the chlorine in an axial position. If we are trying to run a reaction where the Cl needs to be axial, then this effect will speed up the reaction. However, if the Cl is locked in an equatorial position, then the reaction will be too slow:

Now we understand why this can be important. So let's go step by step in determining which of two chair conformations is more stable.

If you have only one group on the ring, then the more stable chair will be the one with the group in an equatorial position:

More stable

If you have two groups, then it is best to put them both in equatorial positions, if possible:

More stable

If only one can be equatorial in either chair, then the more stable chair will be the one with the larger group in the equatorial position:

More stable

In the example above, we have a choice to put the *tert*-butyl group in an equatorial position or the methyl group in an equatorial position. We can't have both. So we put the *tert*-butyl in an equatorial position.

If you have more than two groups, you just use the same logic that we used above to choose the more stable chair. Just try to put the largest groups in equatorial positions.

EXERCISE 6.37 For the following compound, draw the most stable chair conformation:

Answer We begin by drawing both chair conformations (if you have trouble with this, review the two previous sections in this chapter):

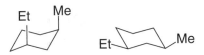

Now we select the chair that has the larger group in the equatorial position. In this case, both groups can be equatorial, so we choose this one:

Et⌐⌐⌐Me

PROBLEMS For each compound below, draw the most stable chair conformation.

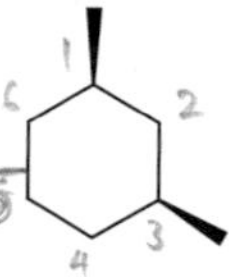

6.38

6.39

6.40

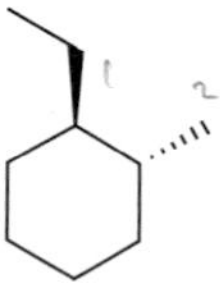

6.41

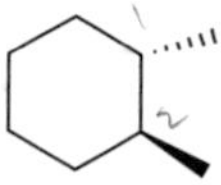

6.42

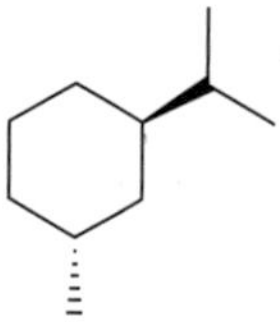

6.43

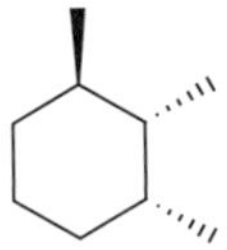

6.44

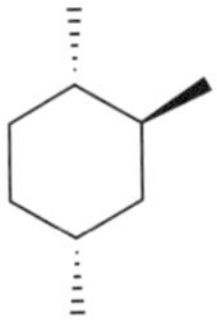

6.45

6.7 DON'T BE CONFUSED BY THE NOMENCLATURE

Some nomenclature is always confusing to students, so it is worth a couple of paragraphs to clear things up. When two groups are both up or both down, we call them *cis* to each other; when one group is up and one group is down, we call them *trans* to each other:

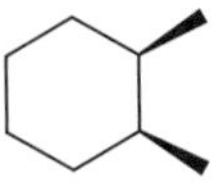

cis 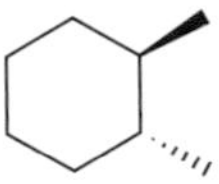   trans

Do not confuse this with cis and trans double bonds. There are no double bonds here. Don't draw any double bonds. It is amazing how many students will draw a double bond when you ask them to draw *cis*-1,2-dimethyl cyclohexane. Remember that the ending –ane– means that there are no double bonds anywhere in the molecule. The only comparison between double bonds and disubstituted cylohexanes is that, in both cases, cis means "on the same side" and trans means "on opposite sides."

CHAPTER 7
CONFIGURATIONS

In the previous chapter, we saw that molecules can assume many different conformations, much like a person. You can move your hands all around: hold them straight up in the air, out to the sides, straight down, and so on. In all of these positions, your right hand is still your right hand, regardless of how you move it around. There is no way to twist your right hand to turn it into a left hand. The reason it is always a right hand has nothing to do with the fact that it is connected to your right shoulder. If you were to chop off your arms and sow them on backward (*don't try this at home!*), you would not look normal. You would look like someone with his right hand attached to his left shoulder and vice versa. You would look very strange, to say the least.

Your right hand is a right hand because it fits into a right-handed glove, and it does not fit into a left-handed glove. This will always be true no matter how you move your hand. Molecules can have this property also.

It is possible for a molecule to have a region where there are two possibilities for how the atoms can be connected in 3D space, much like the difference between a right hand and a left hand. Instead of "right hand" and "left hand," we call the two possibilities R and S. When we talk about the configuration of a compound, we are talking about whether it is R or S. If the arrangement is S, then it will always be S no matter how the arms of the molecule twist about as the molecule moves. In other words, the molecule can move into any conformation it wants, but the configuration will never change. The only way for a configuration to change would be to undergo a chemical reaction.

This explains something we saw in the previous chapter: when drawing chair conformations, we saw that up is always up regardless of which chair you draw. That is because up and down are issues of *configuration,* which does not change when the molecule twists into another conformation.

Don't confuse *conformation* with *configuration.* Students confuse these terms all of the time. *Conformations* are the different positions that a molecule can twist into, but *configuration* is a matter of right-handedness or left-handedness (R or S).

In molecules, the regions that can be R or S are called *stereocenters* (or chiral centers—chiral is Greek for "hand," and we can understand the symbolism there). In this chapter, we will learn how to locate stereocenters, how to draw them properly, how to label them as R or S, and what happens when you have more than one stereocenter in a compound.

This is all *extremely* important stuff. You will understand this as soon as you begin learning reactions. You will see that some reactions convert a stereocenter from R into S (and vice versa), while other reactions will not. To predict what the

134

products of a reaction will be, you absolutely *must* know how to show these stereocenters, and you must understand what is happening to them in a reaction.

7.1 LOCATING STEREOCENTERS

For purposes of this course, we will define a stereocenter as a carbon atom with four different groups on it. For example,

This drawing has a carbon in the center with four different groups: ethyl, methyl, bromine, and chlorine. Therefore, we have a stereocenter. Anytime you have four different groups connected to a carbon atom, there will be two ways to arrange the groups in space (always two; never more and never less). These two arrangements are different configurations:

and

These two compounds are different from each other even though the atoms are connected in the same way. The difference between them comes from their positions in 3D space. Therefore, they are called stereoisomers ("stereo" for space). More specifically, they are called *enantiomers*, because the two compounds are mirror images of each other and they are not superimposable. If we construct models of these two compounds, we see that they are not the same—i.e., they cannot be superimposed.

Notice that we are not looking at just the four *atoms* that are connected to the carbon atom in the middle (which would be Br, Cl, C, and C—and we might think that two of these are the same), but we are looking at the entire groups. In other words, whenever we look at the four groups connected to an atom, we are looking at the entire molecule, no matter how big those groups are. Consider the following example:

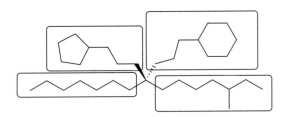

All four of these groups are different.

You must learn how to recognize when an atom has four different groups attached to it. To help you with this, let's begin by seeing the situations that are *not* stereocenters:

Not a stereocenter

The carbon atom indicated above is not a stereocenter because there are two groups that are the same (there are two ethyl groups). The same is true in the following case:

Not a stereocenter

Whether you go around the ring clockwise or counterclockwise, you see the same thing, so this is not a stereocenter. If we wanted to make it a stereocenter, we could do so by putting a group on the ring:

A stereocenter

Usually, stereocenters are drawn with dashes and wedges to show us which configuration is being referred to. If the dashes and wedges are not drawn, then we assume that there is a mixture of equal amounts of both configurations (which we call a *racemic* mixture). In fact, in the compound above, there is a second stereocenter. Can you find it? Each of the two stereocenters in the compound above can be either R or S. Since there are two stereocenters, there will be four possibilities: R,R and R,S and S,R and S,S. Since neither stereocenter has been drawn with dashes and wedges, we must assume that we have all four possible stereoisomers.

EXERCISE 7.1 In the compound below, there is one stereocenter. Find it.

Answer Let's start on the left side and work our way across the compound. The methyl group has three hydrogen atoms, so that can't be it. Then there is a CH_2

group, which also cannot be it, because two groups are the same (two H's). Then you have a carbon with four different groups: ethyl (on the left), methyl (on the right), OH sticking up, and H (don't forget about the H's that are not shown). This is our stereocenter.

PROBLEMS In each of the compounds below, there is one stereocenter. Find it.

7.2

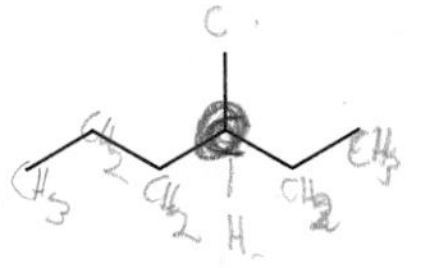

7.3

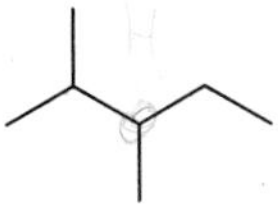

7.4

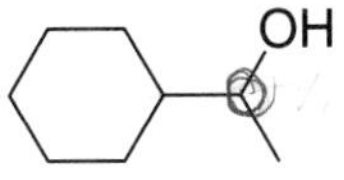

7.5

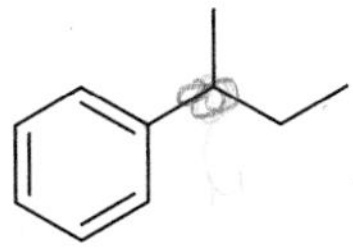

7.6

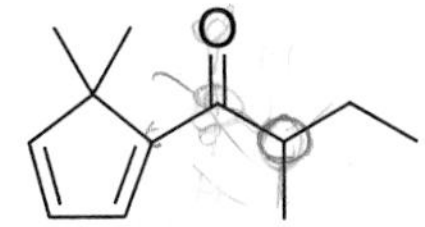

7.7

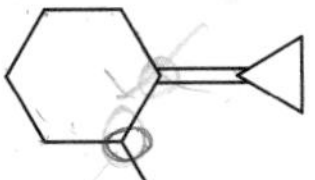

In the previous problems, you knew that you were looking for just one stereocenter. Hopefully, you started to realize some tricks that make it faster to find the stereocenter (for example, ignore CH_2 groups and double bonds). So, now, we will move on to examples where you don't know how many stereocenters there are. There could be five stereocenters or there could be none.

EXERCISE 7.8 In the following compound, find all of the stereocenters, if any:

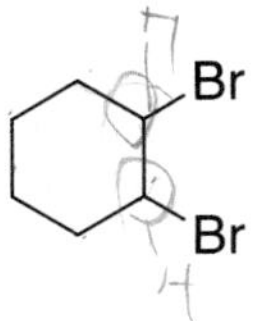

Answer If we go around the ring, we find that there are only six carbon atoms in this compound. Four of them are CH_2 groups, so we know that they are not stereocenters. If we look at the remaining two carbon atoms, we see that each of them is connected to four different groups. They are both stereocenters.

PROBLEMS For each of the compounds below, find all of the stereocenters, if any.

7.9

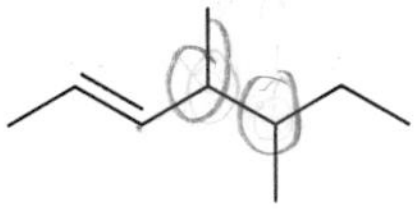

7.10

7.11

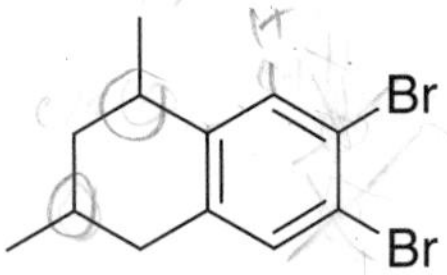

7.12

7.13

7.14

7.15

7.2 DETERMINING THE CONFIGURATION OF A STEREOCENTER

Now that we can find stereocenters, we must now learn how to determine whether a stereocenter is R or S. There are two steps involved in making the determination. First, we give each of the four groups a number (from 1 to 4). Then we use the orientation of these numbers to determine the configuration. So, how do we assign numbers to each of the groups?

We start by making a list of the four *atoms* attached to the stereocenter. Let's look at the following example:

The four atoms attached to the stereocenter are C, C, O, and H. We rank them from 1 to 4 based on atomic weight. To do this, we must either consult a periodic table every time or commit to memory a small part of the periodic table—just those atoms that are most commonly used in organic chemistry:

$$C \quad N \quad O \quad F$$
$$P \quad S \quad Cl$$
$$Br$$
$$I$$

When comparing the four atoms in the example above, we see that oxygen has the highest atomic weight, so we give it the first priority—we give it the number 1. Hydrogen is the smallest atom, so it will always get the number 4 (lowest priority) when a stereocenter has a hydrogen atom. We don't have to worry about what to do if there are two hydrogen atoms, because if there were, it would not be a stereocenter. But it is possible to have two carbon atoms, as in the example above. So how do we figure out which carbon gets the number 2 priority and which gets the number 3 priority?

This is how we rank the two carbon atoms: for each carbon atom, we write a list of three atoms it is connected to (other than the stereocenter). Let's do the example above to see how this works. The carbon on the left side of the stereocenter has four bonds: one to the stereocenter, one to another carbon atom, and then two hydrogen atoms. So, other than the stereocenter, it has three bonds (C, H, and H). Now let's look at the carbon on the right side of the stereocenter. It has four bonds: one to the stereocenter and then three hydrogen atoms. So, other than the stereocenter, it has three bonds (H, H, and H). We compare the two lists and look for the first point of difference:

C	H
H	H
H	H

We see the first point of difference immediately: carbon beats hydrogen. So the left side of the stereocenter gets priority over the right side, and the numbering turns out like this:

EXERCISE 7.16 In the compound below, find the stereocenter, and label the four groups from 1 to 4 using the system of priorities based on atomic weight.

Answer The four atoms attached to the stereocenter are C, C, Cl, and F. Of these, Cl has the highest atomic weight, so its gets the first priority. Then comes F as number 2. We need to decide which carbon atom gets the number 2 and which carbon atom gets the number 3. We do this by listing the three atoms attached to each of them:

Left Side	Right Side
C	C
H	C
H	H

So the right side wins. Therefore, the numbering goes like this:

PROBLEMS In each of the compounds below, find the stereocenter, and label the four groups from 1 to 4 using the system of priorities based on atomic weight.

7.17

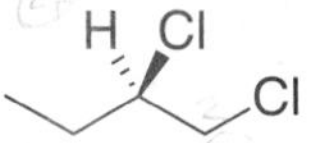

7.18

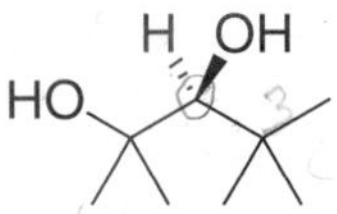

7.19

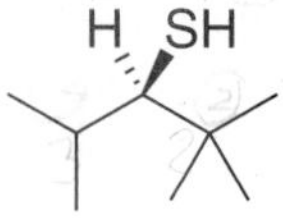

There are a few more situations you should know how to deal with when numbering the four groups. If you are comparing two carbon atoms and you find that the three atoms on one side are the same as the three atoms on the other, then keep going farther out until you find the first difference:

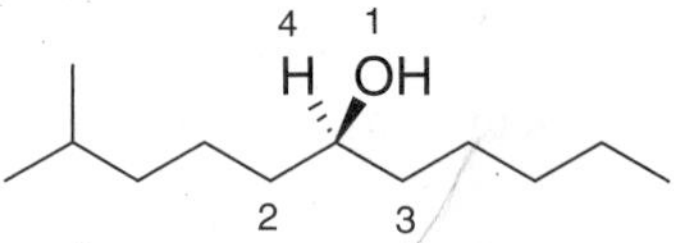

4   1
H OH

2   3

Also, you should know that we are looking for the first point of difference as we travel out, and we don't add the atomic weights. This is best explained with an example:

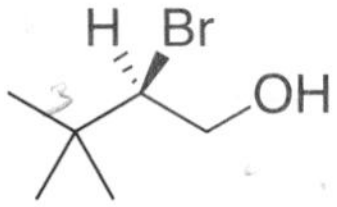

H Br
          OH

In this example, we know that the Br gets the first priority, and the H gets the number 4. When comparing the two carbon atoms, we find the following situation:

Left Side	Right Side
C	O
C	H
C	H

In this case, we do not add the atomic weights and say that the left side wins. Rather we go down the list and compare each row. In the first row above, we have C versus O. That's it, end of story—the O wins. It doesn't matter what comes in the next two rows. Always look for the *first* point of difference. So the priorities go like this:

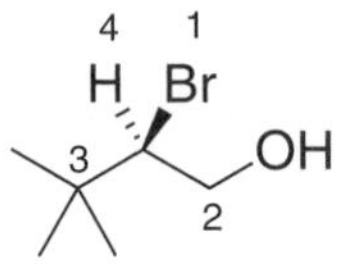

4   1
H Br
3   OH
        2

Finally, you should count a double bond as if the atom were connected to two carbon atoms. For example,

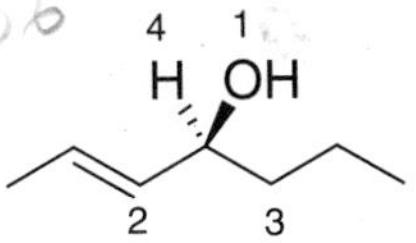

The group on the left gets the number 2, because we counted the following way:

Left Side	Right Side
C	C
C	H
H	H

EXERCISE 7.20 In the compound below, find the stereocenter, and label the four groups from 1 to 4 using the system of priorities based on atomic weight.

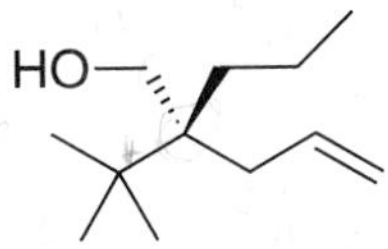

Answer All four atoms connected to the stereocenter are carbon atoms, so we need to compare the three atoms connected in all four cases:

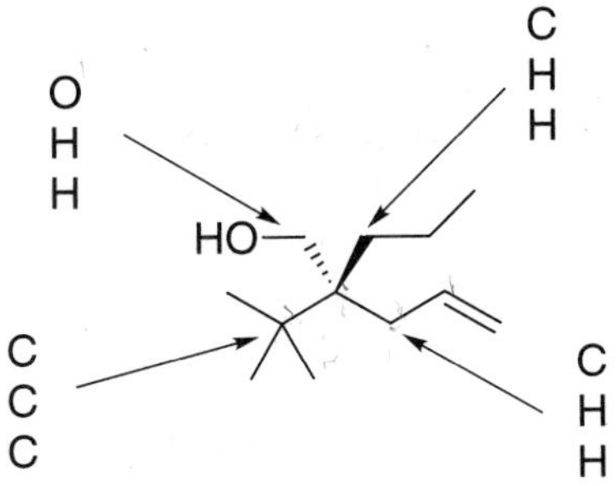

The oxygen wins. Next comes the one with three carbon atoms. The remaining two are the same, so we need to move out one farther on the chain and compare again. Remember to count the double bond like two carbon atoms:

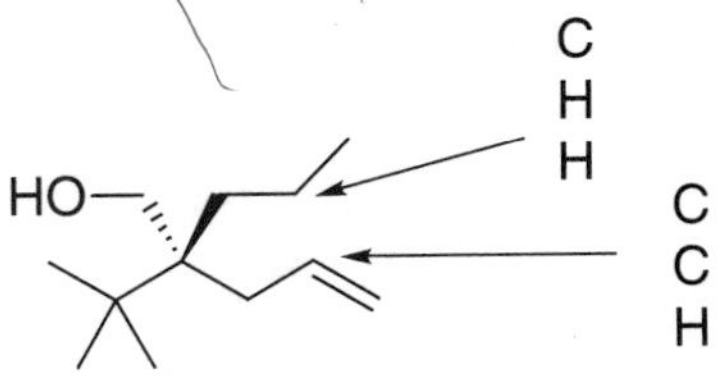

So the order of priorites goes like this:

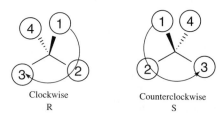

PROBLEMS In each of the compounds below, find the stereocenter and label the four groups from 1 to 4 using the system of priorities based on atomic weight.

7.21

7.22

7.23

7.24

7.25

7.26

Now we need to learn how to use this numbering system to determine the configuration of a stereocenter. The idea is simple, but it is difficult to do if you have a hard time closing your eyes and rotating 3D objects in your mind. For those who cannot do this, don't worry. There is a trick. Let's first see how to do it without the trick.

If the number 4 group is pointing away from us (on the dash), then we ask whether 1, 2, and 3 are going clockwise or counterclockwise:

Clockwise
R

Counterclockwise
S

In the example on the left, we see that 1, 2, 3 go clockwise, which is called R. In the example on the right, we see that 1, 2, 3 go counterclockwise, which is called S. If the molecule is already drawn with the number 4 priority on the dash, then your life is very simple:

The 4 is already on the dash, so you just look at 1, 2, and 3. In this case, they go counterclockwise, so it is S.

It gets a little more difficult when the number 4 is not on a dash, because then you must rotate the molecule in your mind. For example,

Let's redraw just the stereocenter showing the location of the four priorities:

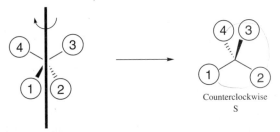

Now we need to rotate the molecule so that the fourth priority is on a dash. To do this, imagine spearing the molecule with a pencil and then rotating the pencil 90°:

Now the 4 is on a dash, so we can look at 1, 2, and 3, and we see that they go counterclockwise. Therefore, the configuration is S.

Let's see one more example:

We redraw just the stereocenter showing the location of the four priorities, and then we spear the molecule with a pencil and rotate 180° to put the 4 on a dash:

Now, the 4 is on a dash, so we can look at 1, 2, and 3, and we see that they go clock-wise. Therefore, the configuration is R.

   And now, for the trick. If you were able to see all of that, great! But if you had trouble seeing the molecules in 3D, there is a simple trick that will help you get the answer every time. To understand how the trick works, you need to realize that if you redraw the molecule so that any two of the four groups are switched, then you have switched the configuration (R turns into S, and S turns into R):

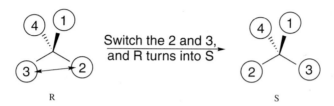

You can switch any two groups and this will happen. You can use this idea to your advantage. Here is the trick: Switch the number 4 with whatever group is on the dash—then your answer is the opposite of what you see. Let's do an example:

This looks tough because the 4 is on a wedge. But let's do the trick: switch the 4 with whichever group is on the dash; in this case, we switch the 4 with the 1:

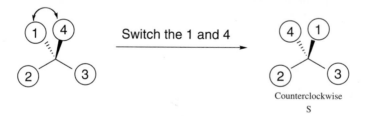

After doing the switch, the 4 is on a dash, and it becomes easy to figure out. It is counterclockwise, which means S. We had to do one switch to make it easy to fig-ure out, which means that we changed the configuration. So if it became S after the switch, then it must have been R before the switch. That's the trick. *But be careful.* This trick will work every time, but you must not forget that the answer you imme-diately get is the opposite of the real answer, because you did one switch.

   Now, let's practice determining R or S when you are given the numbers, so that we can make sure you know how to do this step. You can either visualize the mole-cule in 3D, or you can use the trick—whatever works best for you.

PROBLEMS   In each case, assign the correct configuration (R or S).

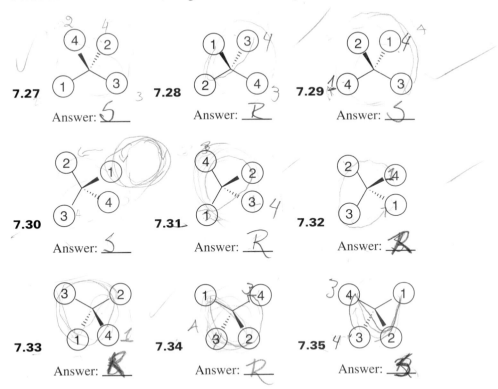

7.27   Answer: _S_

7.28   Answer: _R_

7.29   Answer: _S_

7.30   Answer: _S_

7.31   Answer: _R_

7.32   Answer: _R_

7.33   Answer: _R_

7.34   Answer: _R_

7.35   Answer: _S_

So we know how to assign priorities (1–4), and we know how to use those priorities to determine configuration. Now, let's do some real problems:

EXERCISE 7.36   The compound below has one stereocenter. Find it, and determine whether it is R or S:

OH

Cl   Cl

Answer   The carbon bearing the two chlorine atoms cannot be the stereocenter because there are two Cl atoms (two of the same group). The stereocenter is the carbon atom with the OH group attached. It has four different groups attached to it. Now that we found it, we need to assign priorities.

The O on the dash gets priority number 1, and the hydrogen (not shown, but it is on a wedge) gets number 4. Between the two carbon atoms, the one on the right is connected to two Cl atoms. This wins. So the numbers go like this:

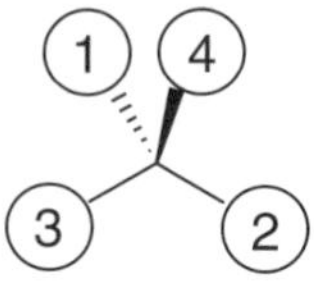

The 4 is on a wedge, which makes the problem a little bit difficult. So let's use the trick. If we switch the 4 and the 1, we get something that is R. So, it must be that it was S before we switched the groups and our answer is S.

PROBLEMS For each compound below, find all stereocenters, and determine their configuration.

7.37

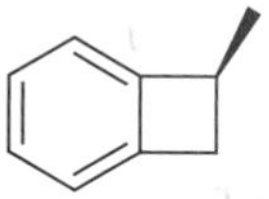

7.38

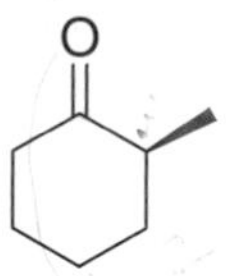

7.39

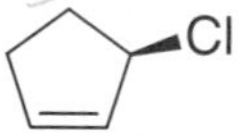

7.40

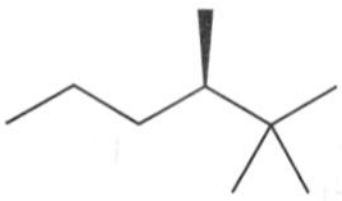

7.41

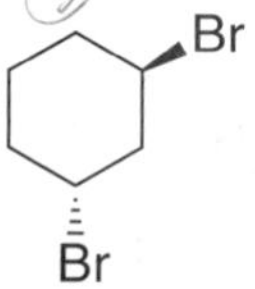

7.42

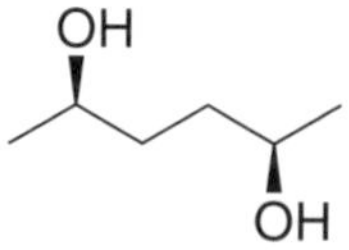

7.3 NOMENCLATURE

When we learned how to name compounds, we said that we would skip over the naming of stereocenters until we learned how to determine configuration. Now that we know how to determine whether a stereocenter is R or S, we can now see how to include this in the name of a compound. It is actually quite simple. If there is a stereocenter in a compound, then you only need to indicate where the stereocenter is and whether it is R or S. Consider the following example:

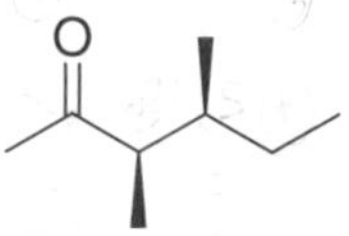

Based on everything we learned in the second chapter (nomenclature), we would name this compound 3,4-dimethylhexan-2-one. Now we must also add the configurations to the name. The stereocenter on the left is R, and the stereocenter on the right is S. We use the numbering system of the parent chain to determine where the stereocenters are. Since the parent chain was numbered left to right, we add (3R,4S) to the name in the section on stereoisomerism:

Stereoisomerism	Substituents	Parent	Unsaturation	Functional group

so the name is: (3R,4S)-3,4-dimethylhexan-2-one. As we saw earlier, we italicize stereoisomerism when it is part of a name.

Now let's turn to a different type of stereoisomerism, one that we already discussed in the chapter on nomenclature. Let's look at double bonds. Recall we indicate the presence of the double bond in a molecule in the unsaturation part of the name using the term –en-:

Stereoisomerism	Substituents	Parent	Unsaturation	Functional group

And we indicated the position of the double bond with the numbering system. But then we saw that there are often two ways for the atoms of a double bond to connect to each other in 3D space. We saw a system for distinguishing these possibilities, using the terms cis and trans:

cis   trans

This was indicated in the first part of the name (stereoisomerism):

Stereoisomerism	Substituents	Parent	Unsaturation	Functional group

This system of identifying double-bond stereoisomers is very limited, because you need two groups that are the same to use the cis/trans system of naming. So what do you do if you have four different groups on a double bond? There are still two possible stereoisomers:

but we cannot use the cis/trans system here. So, we have another system that allows us to differentiate between these two compounds. This other system uses the same numbering based on priorities that we used for stereocenters (based on atomic weights). We look at both sides of the double bond; each side has two groups:

Two groups on this side   Two groups on this side

We begin with one side (let's start with the left), and we ask which of the two groups on the left has priority:

Which of these gets the priority?

The oxygen gets priority over the carbon, based on atomic weights. When comparing the two groups on the right side, the fluorine gets priority over the hydrogen, again based on atomic weights. So now we know which group gets the priority on each side:

In the example above, it was easy to assign priorities, but what about when it gets a little more complicated:

In this example, we have to compare carbon atoms to each other. The groups are all different, so we need to find a way to assign priorities. To do so, we follow the same rules that we did when assigning priorities with R and S:

1. If the atoms are the same on one side, then just move farther out and analyze again.

2. One oxygen beats three carbon atoms (remember to look for the first point of difference)

3. A double bond counts as two individual bonds.

Now that we know how to prioritize, we can use the new system for naming the double bond. Let's go back to our first example. We look at the priority group on one side and the priority group on the other side, and we ask: are these groups on the same side (like cis) or on opposite sides (like trans)?

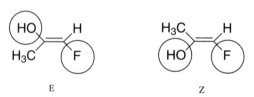

E   Z

The same side is called Z (for the German word "zusammen" meaning together), and opposite sides is called E (for the German word "entgegen" meaning opposite).

This way of naming double bonds is far superior to cis/trans nomenclature because you can use this E/Z system for any double bond, even if all four groups are different. Cis/trans nomenclature requires two groups to be the same.

We include this information in the name of a molecule, very much like we did for R and S configurations. For example, if the double bond is between carbons numbered 5 and 6 on a parent chain, then we would add the term (5E) or (5Z) at the beginning of the name.

EXERCISE 7.43 Name the following compound, including stereochemistry in the beginning of the name.

(3Z,5S)

4-fluro
3,5-dimethyl- hept-3-ene

4-fluro-3,5-dimethyl hept-3-ene

Answer Remember that we go through all five parts of the name, starting at the end and working our way backwards:

Stereoisomerism	Substituents	Parent	Unsaturation	Functional group

We begin with the functional group. There is no functional group (so the suffix is –e). Moving backward in the name, we look for any unsaturation, and there is one double bond (so the unsaturation is –en-). Then, we choose the longest parent that includes the double bond, which is seven carbon atoms long, so the parent is –hept-. There are three substituents (two methyl groups and a fluorine), so we add fluoro and dimethyl before the parent. Then we put in the numbers. We give the double bond the lowest number, so we number from left to right. This gives us

   3,5-dimethyl-4-fluorohept-3-ene

If you do not remember how to do that, then you should review Chapter 5 on nomenclature. Now we are ready to put in the stereochemistry. The double bond is Z:

and there is a stereocenter that is S:

When we number the parent chain, we see that the double bond is at the third carbon in the parent chain, and the stereocenter is at the fifth carbon in the parent chain:

So the name including stereochemistry is

$(3Z,5S)$-4-fluoro-3,5-dimethylhept-3-ene

PROBLEMS Name each of the following compounds. Be sure to include stereochemistry at the beginning of every name:

7.44   Name: _____

7.45   Name: (lR,3R) 3-m   -m el heyl-1- qnol

7.46   Name: Z   D) 3-methyl pent-1-ene

7.47   Name: E) 4-etry-3-mthl hept-2-ene

7.48   Name: E, 4Z)   hex-2.4-ene

7.49   Name: E 42,6Z,   deca-2,4,6,8   terene

yeah.

7.4   DRAWING ENANTIOMERS

We mentioned before that enantiomers are two compounds that are nonsuperimposable mirror images. Let's first clear up the term "enantiomers," since students will often use this word incorrectly in a sentence. Let's compare it to people again. If two boys are born to the same parents, those boys are called brothers. Each one is the brother of the other. If you had to describe both of them, you say that they are brothers. Similarly, when you have two compounds that are nonsuperimposable mirror images, they are called enantiomers. Each one is the enantiomer of the other. Together, they are a pair of enantiomers. But what do we mean by "nonsuperimposable mirror images"? Let's go back to the brother analogy to explain it.

Imagine that the two brothers are twins. They are identical in every way except one. One of them has a mole on his right cheek, and the other has a mole on his left cheek. This allows you to distinguish them from each other. They are mirror images of each other, but they don't look exactly the same (one cannot be superimposed on top of the other). It is very important to be able to see the relationship between different compounds. It is important to be able to draw enantiomers. Later in the course, you will see reactions where a stereocenter is created and both enantiomers are formed. To predict the products, you must be able to draw both enantiomers. In this section, we will see how to draw enantiomers.

The first thing you need to realize is that enantiomers always come in pairs. Remember that they are mirror images of each other. There are only two sides to a mirror, so there can be only two different compounds that have this relationship (nonsuperimposable mirror images). This is very much like the twin brothers. Each brother only has one twin brother, not more.

So we must learn how to draw one enantiomer when we are given the other. When we see the different ways of doing this, we will begin to recognize when compounds are enantiomers and when they are not.

The simplest way to draw an enantiomer is to redraw the carbon skeleton, but invert all stereocenters. In other words, change all dashes into wedges and change all wedges into dashes. For example,

OH

The compound above has a stereocenter (what is the configuration?). If we wanted to draw the enantiomer, we would redraw the compound, but we would turn the wedge into a dash:

This is a pretty simple procedure for drawing enantiomers. It works for compounds with many stereocenters just as easily. For example,

The enantiomer of ⬡ is ⬡

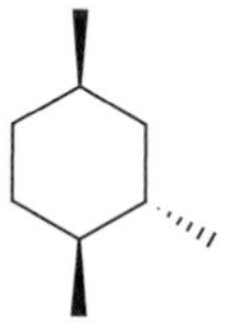

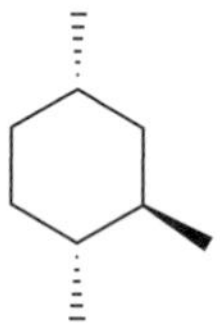

We simply invert all stereocenters. This is actually what we would see if we placed a mirror directly behind the first compound and then looked into the mirror. The carbon skeleton would look the same, but the stereocenters would all be inverted:

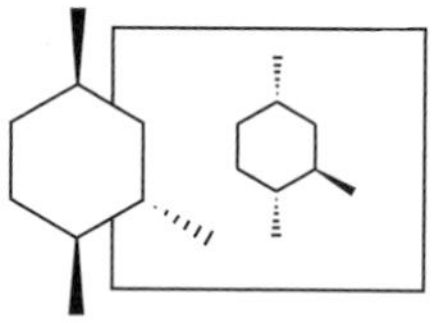

EXERCISE 7.50   Draw the enantiomer of the following compound:

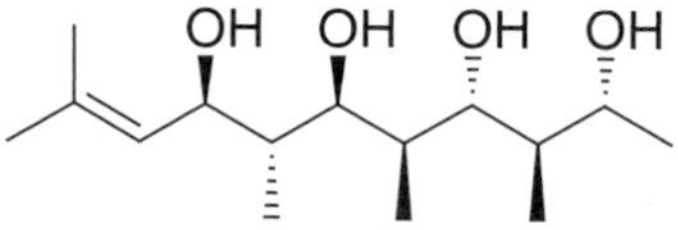

Answer   Redraw the molecule, but invert every stereocenter. Convert all wedges into dashes, and convert all dashes into wedges:

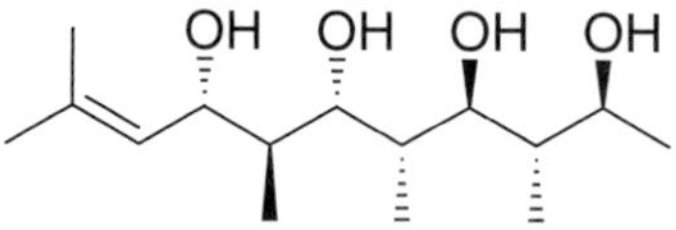

PROBLEMS   Draw the enantiomer of each of the following compounds.

7.51 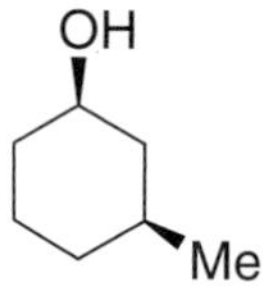   Answer:

7.52 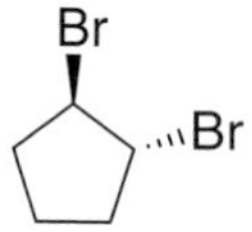   Answer:

7.53 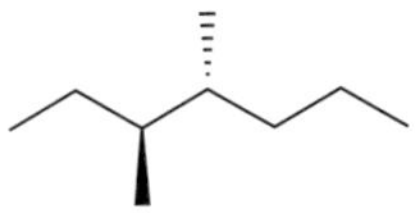   Answer:

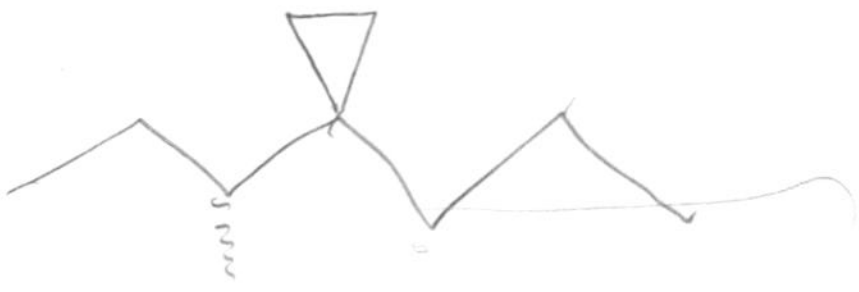

7.54 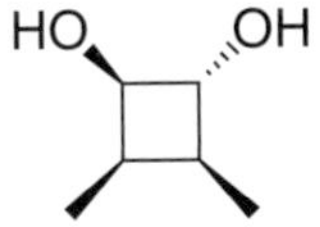   Answer:

7.55 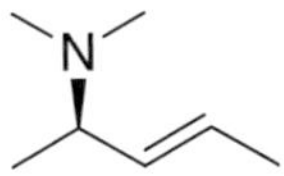   Answer:

7.56 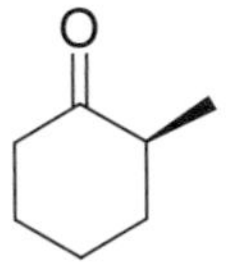   Answer:

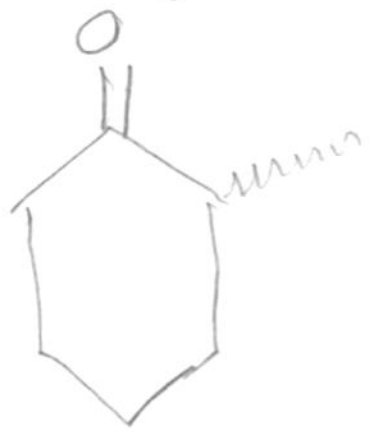

There is another way to draw enantiomers. In the previous method, we placed an imaginary mirror *behind* the compound, and we looked into that mirror to see the reflection. In the second method for drawing enantiomers, we place the imaginary mirror *on the side of* the compound, and we look into the mirror to see the reflection. Let's see an example:

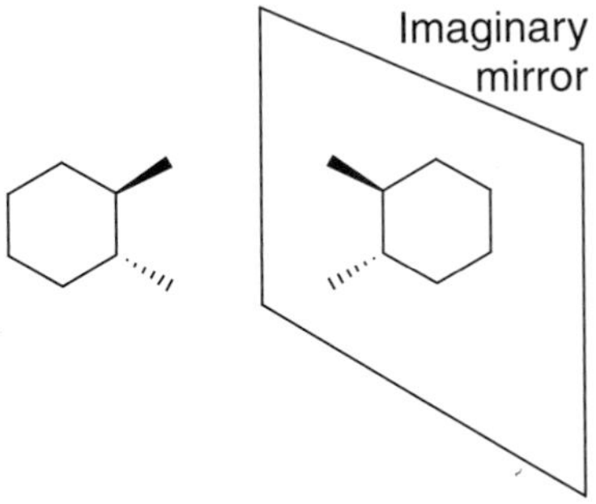

Imaginary mirror

But why do we need a second way of drawing enantiomers? Didn't the first method seem good enough? The first method (switching all dashes with wedges) was pretty simple to do. But there are times when the first method will not work so well. There a few examples of cyclic and bicyclic carbon skeletons where the wedges and dashes are not drawn, because they are implied. We have actually already seen an example of one of these: the chair conformation of substituted cyclohexane.

In this drawing, all of the lines are drawn as straight lines (no wedges and dashes) even though we know that the bonds are not all in the plane of the page. We don't need to draw the wedges and dashes because the geometry can be understood from the drawing. We could try to draw the enantiomer by converting the drawing into a hexagon-style drawing (with wedges and dashes), then drawing the enantiomer using the first method (switching all dashes for wedges), and then redrawing the chair conformation of that enantiomer. But that is a lot of steps to go through when there is a simpler way to draw the enantiomer—just put the imaginary mirror on the side (there is no need to actually draw the mirror), and draw the enantiomer like this:

Whenever we have a structure where the wedges and dashes are implied but not drawn, it is much easier to use this method. There are other examples of carbon skeletons that, by convention, do not show the wedges and dashes. Most of these examples are rigid bicyclic systems. For example,

When dealing with these kinds of compounds, it is much easier to use the second method to draw enantiomers. Of course, if you like this method, you can always use this second method for all compounds (even those that show wedges and dashes).

You should get practice placing the mirror on either side (and you should notice that you get the same result whether you put the mirror on the left side or the right side).

EXERCISE 7.57 Draw the enantiomer of the following compound:

Answer This is a rigid bicyclic system, and the dashes and wedges are not shown. Therefore, we will use the second method for drawing enantiomers. We will place the mirror on the side of the compound, and draw what would appear in the mirror:

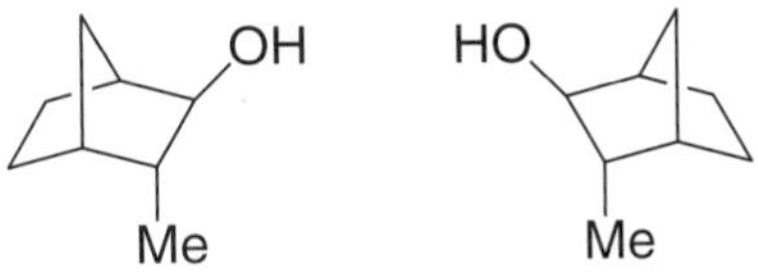

PROBLEMS Draw the enantiomer of each of the following compounds.

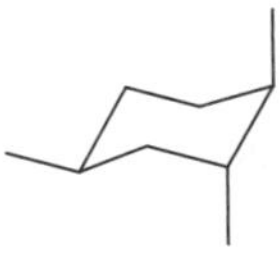

7.58   Answer: _____

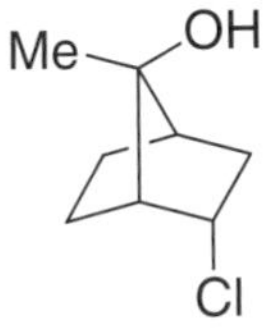

7.59   Answer: _____

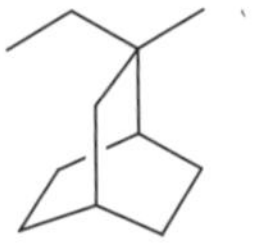

7.60   Answer: _____

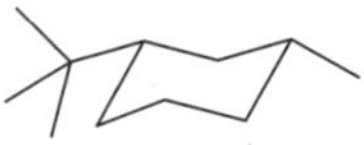

7.61   Answer: _____

7.62   Answer: _____

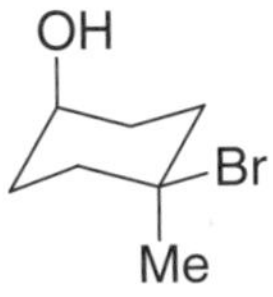

7.63   Answer: _____

7.5 DIASTEREOMERS

In all of our examples so far, we have been comparing two compounds that are mirror images. For them to be mirror images, they need to have different configurations for every single stereocenter. Remember that our first method for drawing enantiomers was to switch all wedges with dashes. For the two compounds to be enantiomers, every stereocenter had to be inverted. But what happens if we have many stereocenters and we only invert some of them?

Let's start off with a simple case where we only have two stereocenters. Consider the two compounds below:

We can clearly see that they are not the same compound. In other words, they are nonsuperimposable. But, they are *not* mirror images of each other. The top stereocenter has the same configuration in both compounds. If they are not mirror images, then they are not enantiomers. So what is their relationship? They are called diastereomers. Diastereomers are any compounds that are nonsuperimposable stereoisomers that are not mirror images of each other.

We use the term "diastereomer" very much like we used the term "enantiomer" (remember the brother analogy). One compound is called the diastereomer of the other, and you can have a group of diastereomers. When we were talking about enantiomers, we saw that they always come in pairs, never more than two. But diastereomers can form a much larger family. We can have 100 compounds that are all diastereomers of each other (if there are enough stereocenters to allow for that many permutations of the stereocenters).

E/Z isomers (or cis/trans isomers) fall under this category. They are called diastereomers, because they are stereoisomers that are not mirror images of each other:

If you are given two stereoisomers, you should be able to determine whether they are enantiomers or diastereomers. All you need to look at are the stereocenters. They must all be of different configuration for the compounds to be enantiomers.

EXERCISE 7.64 Identify whether the two compounds shown below are enantiomers or diastereomers:

Answer There are two stereocenters in each compound. The configurations are different for both stereocenters, so these compounds are enantiomers. In fact, if you were given the first compound only, you could have drawn the enantiomer by using the first method—switching all wedges and dashes.

PROBLEMS   For each pair compounds below, determine whether the pair are enantiomers or diastereomers.

7.65

Answer: _____

7.66

Answer: _____

7.67

Answer: _____

7.68

Answer: _____

7.69

Answer: _____

7.70

Answer: _____

7.6   MESO COMPOUNDS

This is a topic that notoriously confuses students, so let's start off with analogy. Let's use the analogy of the twin brothers who look identical except for one feature: one of them has a mole on the left side of his face, and the other has a mole on the right side of his face. You can tell them apart based on the mole, and the brothers are mirror images of each other. Imagine that their parents had other sets of twins, lots of sets of twins. So, all in all, there are a lot of siblings (who are all brothers and sisters of each other), but they are paired up, 2 in a group. Each child has *only one* twin sibling, who is his or her mirror image. Now imagine that the parents, out of nowhere, have a one more child who is born without a twin—just a regular one-baby birth. When you look at this family, you would see a lot of sets of twins, and then one child who has no twin (and has two moles—one on each side of his face). You might ask that child, where is your twin? Where is your mirror image? He would answer: I don't have a twin. I am the mirror image of myself. That's why the family has an odd number of children, instead of an even number.

The analogy goes like this: when you have a lot of stereocenters in a compound, there will be many stereoisomers (brothers and sisters). But, they will be paired up into sets of enantiomers (twins). Any one molecule will have many, many diastereomers (brothers and sisters), but it will have only one enantiomer (its mirror image twin). For example, consider the following compound:

This compound has five stereocenters, so it will have many diastereomers (compounds where only some of the wedges have been inverted). There are many, many possible compounds that fit that description, so this compound will have many brothers and sisters. But this compound will only have one twin—only one enantiomer (there is *only one* mirror image of the compound above):

It is possible for a compound to be its own mirror image. In such a case, the compound will not have a twin. It will be all by itself, and the total number of stereoisomers will be an odd number, rather than an even number. That one lonely compound is called a meso compound. If you try to draw the enantiomer (using either of the methods we have seen), you will find that your drawing will be the same compound as what you started with.

So how do you know if you have a meso compound?

A meso compound has stereocenters, but the compound also has symmetry that allows it to be the mirror image of itself. Consider *cis*-1,2-dimethylcyclohexane as an example. This molecule has a plane of symmetry cutting the molecule in half. Everything on the left side of the plane is mirrored by everything on the right side:

If a molecule has an internal plane of symmetry, then it is a meso compound. If you try to draw the enantiomer (using either one of the two methods we saw), you will find that you are drawing the same thing again. This molecule does not have a twin. It is its own mirror image:

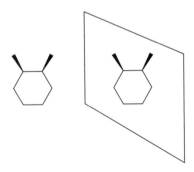

So, to be meso, the compound needs to be the same as its mirror image. We have seen that this can happen when we have an internal plane of symmetry. It can also happen when the compound has a center of inversion. For example,

This compound does not possess a plane of symmetry, but it does have a center of inversion. If we invert everything around the center of the molecule, we regenerate the same thing. Therefore, this compound will be superimposable on its mirror image, and the compound is meso. You will rarely see an example like this one, but it is not correct to say that the plane of symmetry is the only symmetry element that makes a compound meso. In fact, there is a whole class of symmetry elements (to which the plane of symmetry and center of inversion belong) called Sn axes, but we will not get into this, because it is beyond the scope of the course. For our purposes, it is enough to look for planes of symmetry.

There is one fail-safe way to tell if a compound is a meso compound: simply draw what you think should be the enantiomer and see if you can rotate the new drawing in any way to superimpose on the original drawing. If you can, then the compound will be meso. If not, then your second drawing is the enantiomer of the original compound.

EXERCISE 7.71 Is the following a meso compound?

Answer We need to try to draw the mirror image and see if it is just the same compound redrawn. If we use the second method for drawing enantiomers (placing the mirror on the side), then we will be able to see that the compound we would draw is the same thing:

Therefore, it is a meso compound.

A simpler way to draw the conclusion would be to recognize that the molecule has an internal plane of symmetry that chops through the center of one of the methyl groups:

PROBLEMS Identify which of the following compounds is a meso compound.

7.72   7.73   7.74

7.7 DRAWING FISCHER PROJECTIONS

There is an entirely different way to draw stereocenters (instead of using regular bond-line drawings with dashes and wedges). Fischer projections are helpful for drawing molecules that have many stereocenters, one after another. These drawings look like this:

2 stereocenters   3 stereocenters   4 stereocenters

First we need to understand exactly what these drawings mean, and then we will learn a step-by-step method for drawing them properly.

Using Fischer projections saves time because we don't have to draw all of the dashes and wedges for each stereocenter. Instead, we draw only straight lines, with the idea that all horizontal lines are coming out at us and all vertical lines are going away from us. Let's see exactly how this works. Consider the following molecule, which is drawn in a zigzag format (R_1 and R_2 represent groups whose identities are not being defined yet, because it does not matter for now):

Remember that all of the single bonds are all freely rotating, so there are a large number of conformations that the molecule can assume. When we rotate a single bond, the dashes and wedges change, but this is *not* because the configuration has changed. Remember that configurations do not change when a molecule twists and bends. Watch what happens when we rotate one of the single bonds:

Notice that R_1 is now pointing straight down, and the OH is now on a dash. The configuration has *not* changed. If you need to convince yourself of this, determine the configuration of that stereocenter in each of two drawings. You will see that it has not changed.

Now let's draw another of the possible conformations for this molecule. If we rotate a couple more single bonds until the carbon skeleton is looping around like a bracelet, we get the following conformation:

The molecule is twisting and bending around all of the time, and the conformation with the bracelet-shaped skeleton is just one of the possible conformations. The molecule probably spends very little of its time like this (just from a statistical point of view—this is one of many possible conformations), but this is the conformation that we will use to draw our Fischer projection.

Now imagine piercing a pencil through R_1 and R_2 (the pencil is represented by the dotted line below). If you grab the ends of the pencil and rotate, you will find that

R_1 and R_2 will stay in the page, but the rest of the molecule will pop out in front of the page:

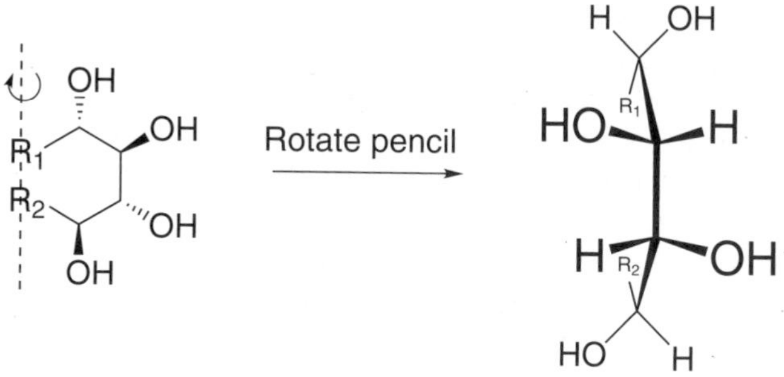

Now we imagine flattening the skeleton into a straight vertical line, and we re-draw the molecule using only straight lines for the groups:

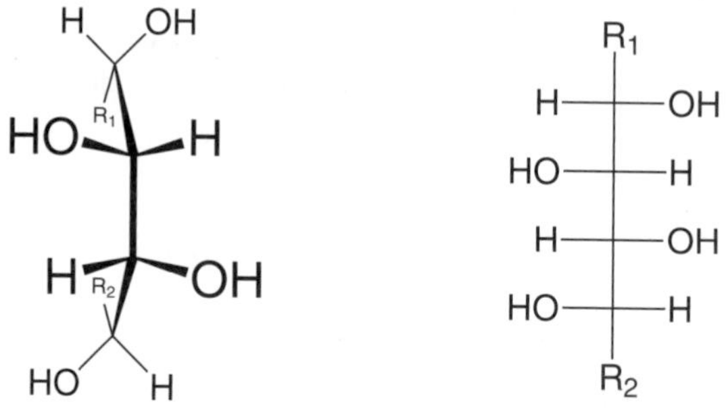

This is our Fischer projection. All of the configurations can be seen on this drawing, because we are able to picture in our minds what the 3D shape is. So the rule is that all horizontal lines are coming out at us, and all vertical lines are going away from us:

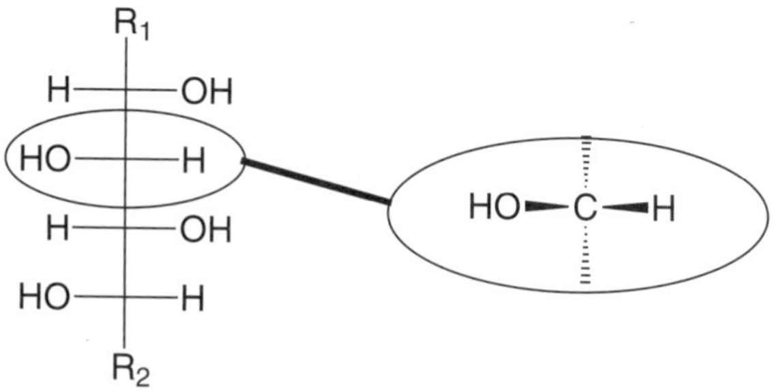

You might be wondering how you would determine the configuration of a stereocenter when you are given a Fischer projection. If each stereocenter is drawn as two wedges and two dashes, how do you figure out how to look at the stereo-center? The answer is simple. Just choose one dash and one wedge, and draw them as straight lines. It doesn't matter which ones you choose—you will get the answer right regardless:

$$\underset{\text{CH}_2\text{CH}_3}{\overset{\text{CH}_3}{\text{HO}+\text{H}}} = \underset{\text{CH}_2\text{CH}_3}{\overset{\text{CH}_3}{\text{HO}-\text{C}-\text{H}}} = \underset{\text{CH}_2\text{CH}_3}{\overset{\text{CH}_3}{\text{HO}-\text{C}-\text{H}}} \quad \text{or} \quad \underset{\text{CH}_2\text{CH}_3}{\overset{\text{CH}_3}{\text{HO}-\text{C}-\text{H}}} \quad \text{etc.}$$

Once you have a drawing with two straight lines, one dash and one wedge, then you should be able to determine whether the stereocenter is R or S. If you cannot, then you should go back and review the section on assigning configuration.

Fischer projections can also be used for compounds with just one stereocenter, as above, but they are usually used to show compounds with multiple stereocenters. You will utilize Fischer projection heavily when you learn about carbohydrates at the end of the course.

Now we can understand why we cannot draw a Fischer projection sideways. If we did, we would be inverting the stereocenter. To draw the enantiomer of a Fischer projection, do not turn the drawing sideways. Instead, you should use the second method we saw for drawing enantiomers (place the mirror on the side of the compound and draw the reflection). Recall that this was the method that we used for drawings where wedges and dashes were implied but not shown. Fischer projections are another example of drawings that fit this criterion:

$$\begin{array}{cc}
\text{COOH} & \text{COOH} \\
\text{HO}+\text{H} & \text{H}+\text{OH} \\
\text{HO}+\text{H} & \text{H}+\text{OH} \\
\text{H}+\text{OH} & \text{HO}+\text{H} \\
\text{HO}+\text{H} & \text{H}+\text{OH} \\
\text{CH}_2\text{OH} & \text{CH}_2\text{OH}
\end{array}$$

Enantiomers

EXERCISE 7.75 Determine the configuration of the stereocenter below. Then draw the enantiomer.

$$\underset{\text{Me}}{\overset{\text{CH}_2\text{OH}}{\text{H}+\text{Cl}}}$$

Answer We begin by drawing the stereocenter as it is implied by the Fischer projection:

$$\underset{\text{Me}}{\overset{\text{CH}_2\text{OH}}{\text{H}-\text{C}-\text{Cl}}}$$

Next, we choose one dash and one wedge, and we turn them into straight lines (it doesn't matter which dash or which wedge we choose):

$$CH_2OH$$
$$H-C\blacktriangleleft Cl$$
$$Me$$

Then we assign priorities based on atomic weight:

②
$$CH_2OH$$
④ H−C◀Cl ①
$$Me$$
③

The 4 is not on a dash, so we switch it with the 3 so it can be on a dash, and we see that the configuration is S. Since we had to do a switch to get this, the configuration of the original stereocenter (before the switch) was R:

②
③—C◀①
④ **S**

therefore:

②
④—C◀①
③ **R**

Now, we need to draw the enantiomer. For Fischer projections, we use the method where we place a mirror on the side, and then we draw the reflection:

$$CH_2OH$$
H——Cl
$$Me$$

$$CH_2OH$$
Cl——H
$$Me$$
Enantiomer

PROBLEMS For each compound below, determine the configuration of the stereocenter, and then draw the enantiomer.

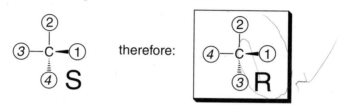

7.76
Et
H——Br
Me

7.77
$$CH_2OH$$
Me——Br
Et

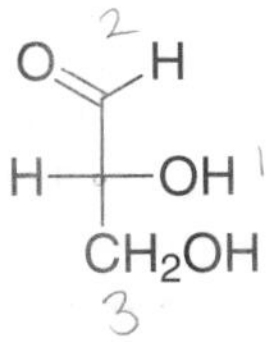

7.78

COOH structure with:
O=C-H (2)
4 H—OH (1)
CH₂OH (3)

[handwritten annotations: R, O=C(H)(OH), —H, CH₂OH]

PROBLEMS For each compound below, determine the configuration of every stereocenter. Then draw the enantiomer of each compound below (the COOH group is a carboxylic acid group).

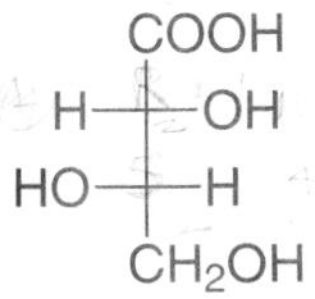

7.79

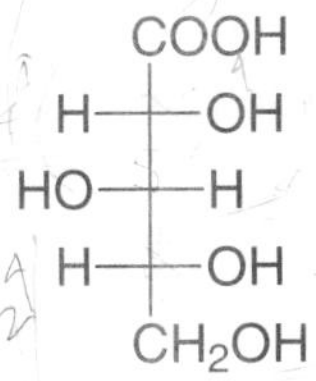

COOH
H——OH
HO——H
CH₂OH

7.80

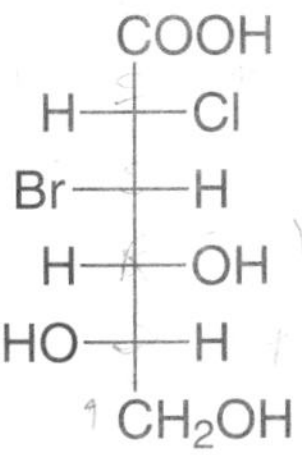

COOH
H——OH
HO——H
H——OH
CH₂OH

7.81

COOH
H——Cl
Br——H
H——OH
HO——H
CH₂OH

7.8 OPTICAL ACTIVITY

Students confuse R/S with +/− all of the time, so let's conclude our chapter by clearing up the difference. Compounds are chiral if they have stereocenters and they are not meso compounds. A chiral compound will have an enantiomer (a nonsuperimposable mirror image). An interesting thing happens when you take a chiral compound and subject it to plane-polarized light. The plane of the polarized light rotates as it passes through the sample. If this rotation is clockwise, then we say the rotation is +. If the rotation is counterclockwise, then we say the rotation is −. If we want to refer to a racemic mixture (an equal mixture of both enantiomers), we will often say (+/−) in the beginning of the name to mean that both enantiomers are present in solution (and the rotations cancel each other).

Do not confuse clockwise rotation of plane-polarized light with clockwise ordering of atomic weights when determining configurations. They are not related. When determining configuration, we impose a set of man-made rules to help us distinguish between the two possible configurations. By using these rules, we will always be able to communicate which configuration we are referring to, and we only need one letter to communicate this (R or S) if we use the rules properly. However, $+/-$ is totally different.

The rotation of plane polarized light (either $+$ or $-$) is not a man-made convention. It is a physical effect that is measured in the lab. It is impossible to predict whether a compound will be $+$ or $-$ without actually going into the lab and trying. If a stereocenter is R, this does not mean that the compound will be $+$. It could just as easily be $-$. In fact, whether a compound is $+$ or $-$ will depend on temperature. So a compound can be $+$ at one temperature and $-$ at another temperature. But clearly, temperature has nothing to do with R and S. So, don't confuse R/S with $+/-$. They are totally independent and unrelated concepts.

You will never be expected to look at a compound that you have never seen and then predict in which direction it will rotate plane-polarized light (unless you know how the enantiomer rotates plane-polarized light, because enantiomers have opposite effects). But you will be expected to assign configurations (R and S) for stereocenters in compounds that you have never seen.

CHAPTER 8

MECHANISMS

Mechanisms are your key to success in this course. If you can master the mechanisms, you will do very well in this class. If you don't master mechanisms, you will do poorly in this class. What are mechanisms and why are they so important?

When two compounds react with each other to form new and different products, we try to understand *how* the reaction occurred. Every reaction involves the flow of electron density—electrons move to break bonds and form new bonds. Mechanisms illustrate how the electrons move during a reaction. The flow of electrons is shown with curved arrows; for example,

These arrows show us how the reaction took place. For most of the reactions that you will see this semester, the mechanisms are well understood (although there are some reactions whose mechanisms are still being debated today). You should think of a mechanism as "bookkeeping of electrons." Just as an accountant will do the bookkeeping of a company's cash flow (money coming in and money going out), the mechanism of a reaction is the bookkeeping of the flow of electrons.

When you understand a mechanism, you will understand why the reaction took place, why the stereocenters turned out the way they did, and so on. If you do not understand the mechanism, then you will find yourself memorizing the exact details of every single reaction. Unless you have a photographic memory, that will be a very difficult challenge. By understanding mechanisms, you will be able to make more sense of the course content, and you will be able to better organize all of the reactions in your mind.

The mechanisms that you will learn in the first half of your course are the most critical ones. This is the time when you will either master arrow pushing and mechanisms or you will not master them. If you don't, you will struggle with all mechanisms in the rest of the course, which will turn your organic chemistry experience into a nightmare. It is absolutely critical that you master the mechanisms for the early reactions that you cover. That way, you will have the tools that you need to understand all of the other mechanisms in your course.

In this chapter, we will *not* learn every mechanism that you need to know. Rather, we will focus on the tools that you need to properly read a mechanism and abstract the important information. You will learn some of the basic ideas behind arrow pushing in mechanisms, and these ideas will help you conquer the early mechanisms that you will learn. The second half of this chapter provides a place for you to keep a list of mechanisms as you progress through the course. This list (which you will fill out as you go along) is arranged so that you will have the key information at your fingertips, and you will be able to use the list as a study guide for your exams.

8.1 CURVED ARROWS

We have already gotten quite a bit of experience with curved arrows in chapter 2 (Resonance). There is one very major difference between curved arrows for drawing resonance structures and curved arrows for drawing mechanisms. With resonance structures, we saw that the electrons were not really moving at all. We were pretending that they were moving so that we could draw all of the resonance structures. By contrast, the curved arrows that we use in mechanisms refer to the actual *movement of electrons*. Electrons are moving to break and form bonds (hence the term *chemical reaction*). Why are we stressing this difference? We first need to understand what arrows represent before we can move on to the rules of pushing arrows.

When we learned how to draw resonance structures, we saw two commandments that we must not violate: (1) never break a single bond, and (2) never violate the octet rule. When drawing mechanisms, we are trying to understand where the electrons actually moved to break and form bonds. Therefore, it is OK to break single bonds. In fact, it happens in almost every reaction. So when drawing mechanisms there is only one commandment to follow: never violate the octet rule.

Now that we have some of the ground rules down, let's just have a quick review of curved arrows, and the different types of arrows that you can draw. Every curved arrow has a *head* and a *tail*. It is essential that the head and tail of every arrow be drawn in precisely the proper place. *The tail shows where the electrons are coming from, and the head shows where the electrons are going:*

Tail   Head

Therefore, there are only two things that you have to get right when drawing each arrow. The tail needs to be in the right place and the head needs to be in the right place. Remember that electrons exist in orbitals, either as lone pairs or as bonds. So the *tail* of an arrow can only come *from a bond* or *from a lone pair:* The *head* of an arrow can only be drawn to *make a bond* or to *make a lone pair:* In total, this gives us four possibilities:

1. Lone pair → bond
2. Bond → lone pair
3. Bond → bond
4. Lone pair → lone pair

The last possibility does not work, because we cannot push electrons from one lone pair to another (at least not in one step). So we only have to consider the first three possibilities. Every arrow you will see will belong to one of these three categories, so let's see examples of each of the three categories.

From a Lone Pair to a Bond

Consider the second step in the mechanism above, where we are forming a single bond:

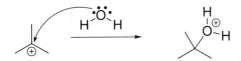

The tail of the arrow is coming from a lone pair on the oxygen atom, and the head of the arrow is going to form a bond between oxygen and carbon. Since the head of the arrow is placed on an atom, it might *seem like* the electrons are going from a lone pair to a lone pair, but they are not. The electrons are going from the oxygen lone pair to form a bond to the carbon atom. If this makes you unhappy, there is an alternative way of drawing the arrow that shows it more clearly:

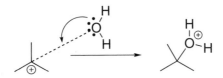

The dotted line shows the bond that is about to form, and we draw the arrow to that dotted line. In this drawing it is very clear that the head of the arrow is going to form a bond. When you see an arrow drawn the first way (where it looks like it is going to an atom rather than to form a bond), don't be confused by this—it really is just going to form a bond.

From a Bond to a Lone Pair

Consider the first step of the mechanism above, where we are breaking a single bond:

The tail of the arrow is on a bond, and the head of an arrow is forming a lone pair on the chlorine atom. The two electrons of the bond used to be shared between the carbon and the chlorine atoms. But now, both electrons are going on the chlorine. So the carbon has lost an electron, and the chlorine has gained one. This is why the carbon ends up with a positive charge, and the chlorine gets a negative charge.

By the way, a chlorine atom with a negative charge is called a chloride ion (-ide- implies the negative charge). So in this reaction chloride is popping off of the molecule to form a carbocation (a carbon with a positive charge).

From a Bond to a Bond

Consider the first arrow in the example below, where we are using the electrons of the pi bond to attack a proton (H^+), and kicking off Cl^- in the process:

The first arrow has its tail on the pi bond, and the head is being used to form a bond between a carbon atom and the proton.

You will notice in the example above that there are two arrows. The first arrow is going from a bond to a bond. But the second arrow is going from a bond to form a lone pair. So we see that you can have more than one type of arrow together in one step of a mechanism.

In fact, it is possible to have all three types of arrows in one step of a mechanism. Consider the example below:

Notice that there is one long flow of electron density, illustrated with three arrows. We begin at the tail of the arrow on the base, because that is where the flow starts. This arrow is going from a lone pair to form a bond. The second arrow goes from a bond to form a bond, and the third arrow goes from a bond to form a lone pair on X. This type of reaction is called an elimination reaction, because we are *eliminating* H^+ and X^- to form a double bond:

Notice that the arrows are all going from one end of the molecule to the other. *Never* draw arrows going in opposite directions. That would not make any sense! To see what we mean by this, consider the example below:

This type of reaction will be covered much later on in your course, but let's use it now as an example. Notice that there are two steps to this mechanism. In the first step, we have two arrows: from a lone pair to form a bond, and then from a bond to form a lone pair:

Bond to lone pair

Lone pair to bond

In the second step of the mechanism, we also have two arrows: from a lone pair to form a bond, and then from a bond to form a lone pair:

Lone pair to bond

Bond to lone pair

If we consider the overall reaction, we notice that the OH$^-$ is replacing the Cl. If we look at how the electrons flowed, we see that it all started at the negative charge of the attacking OH$^-$. This charge flowed up temporarily on to the oxygen atom of the C=O in step 1 of the mechanism, and then the charge flowed back down to kick off Cl$^-$:

Electron flow up

Electrons flow
back down

When we consider how the charge flowed throughout the whole reaction, it might be tempting to draw it all in one step, like this:

However, this is no good, because we have two arrows going in opposite directions:

Never draw arrows in opposite directions. That would imply that the electrons were flowing in opposite directions *at the same time*. That is not possible. In this reaction, the electrons first flowed up, and then they flowed back down. So we have to draw it as two steps:

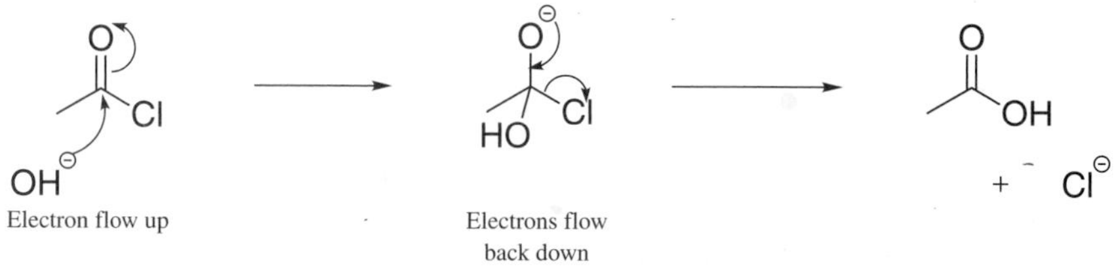

Electron flow up   Electrons flow
                                    back down

Before we can practice drawing arrows, we first need to make sure that we can identify the three different arrow types. This is important, because it will get you accustomed to the types of arrows that are acceptable to draw.

EXERCISE 8.1   In the example below, classify each arrow that you see into one of the following three types:

1. Bond → bond
2. Bond → lone pair
3. Or lone pair → bond

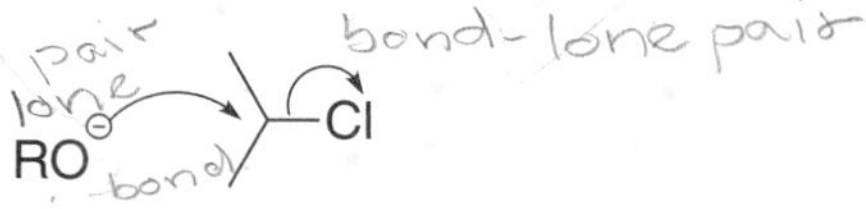

Answer   The first arrow is going from a lone pair on the oxygen to form a bond between the oxygen and carbon. So, this arrow is of the type lone pair → bond.

The second arrow is going from a bond to form a lone pair, so the second arrow is of the type bond → lone pair.

PROBLEMS   For each of the following examples, classify each arrow that you see into one of the three types that we discussed.

8.2

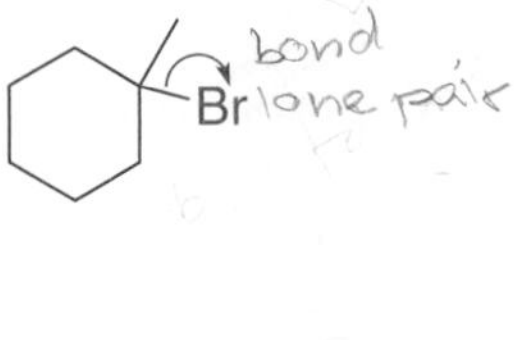

8.3

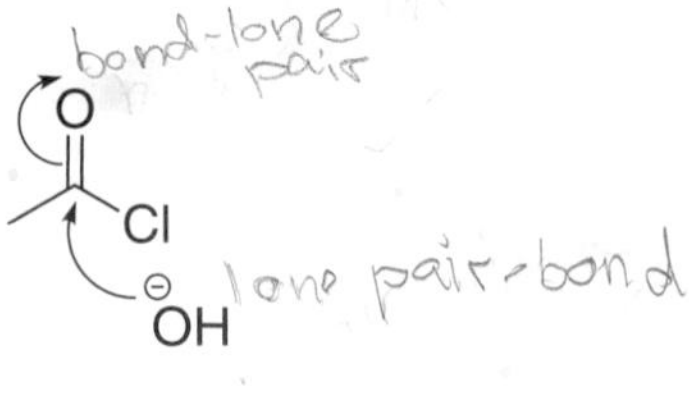

8.4

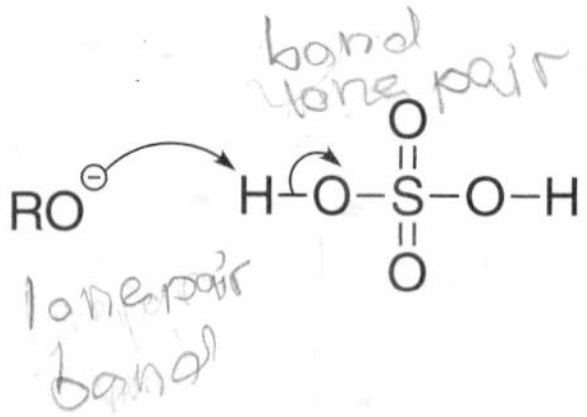

8.5

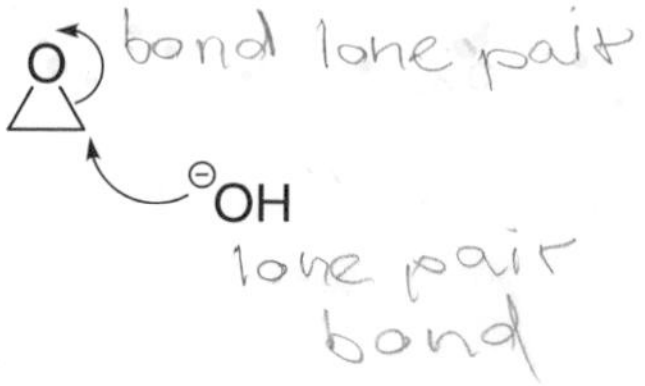

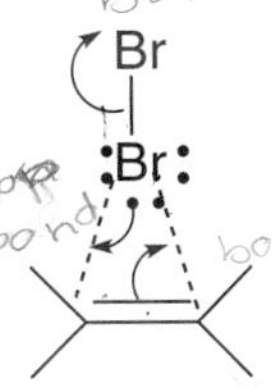

8.6 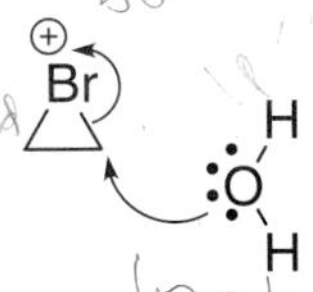   8.7

8.2 ARROW PUSHING

Now that we know what kinds of arrows are acceptable, we can begin to practice drawing them (or "pushing" them, as its called). To do this, we need to learn how to analyze a step in a mechanism, and train our eyes to look for all of the lone pairs and all of the bonds. We have said that all arrows are coming from or going to either lone pairs or bonds. So it makes sense that we need to be able to look at a step in a mechanism and determine which bonds have changed and which lone pairs have changed. Let's see this in an example.

EXERCISE 8.8 Complete the mechanism of the following reaction by drawing the proper arrows in each step:

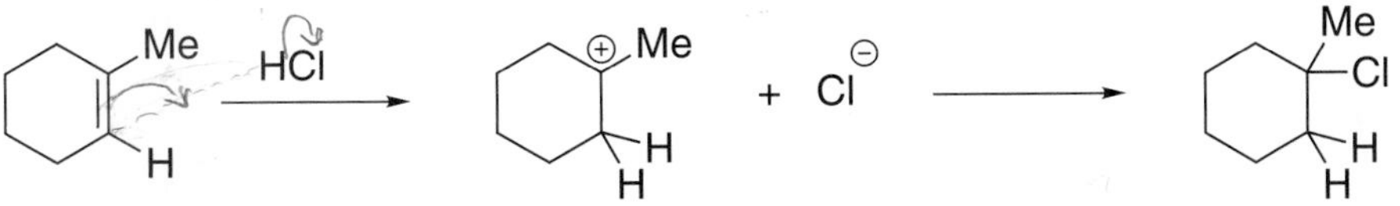

Answer We need to look for all changes for bonds or lone pairs. In the first step, the double bond is disappearing, one of the carbon atoms of the double bond is forming a new bond to a proton (H^+), and we are breaking the H—Cl bond to kick off Cl^-. So we have broken two bonds (C=C, and H—Cl) and we have formed one bond (C—H) and one extra lone pair (on Cl). Therefore, we will need two arrows to make this happen. Where do we start?

Keep in mind that electron density always flows in one direction. In this example we can see which direction the flow went, because in the end we had a positive charge on one side and a negative charge developed on the chlorine. We can use that information to figure out the direction of the flow. The first arrow needs to show the double bond going to form a bond to the proton (from a bond to a bond) and then we need a second arrow to show the bond from the H—Cl going to form a negative charge on the chlorine atom (from a bond to a lone pair):

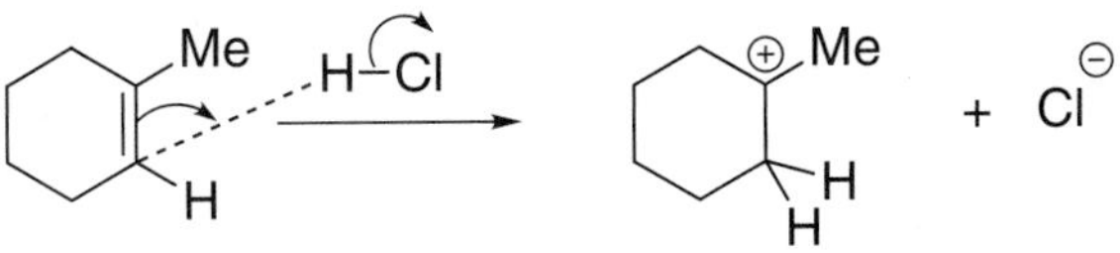

In the next step, again we look for all changes to lone pairs or bonds. We see that the Cl is giving up one of its lone pairs to form a bond with a carbon (C$^+$). So, we need only one arrow, from a lone pair to form a bond:

PROBLEMS   For each transformation below, complete the mechanism by drawing the proper arrows.

8.9

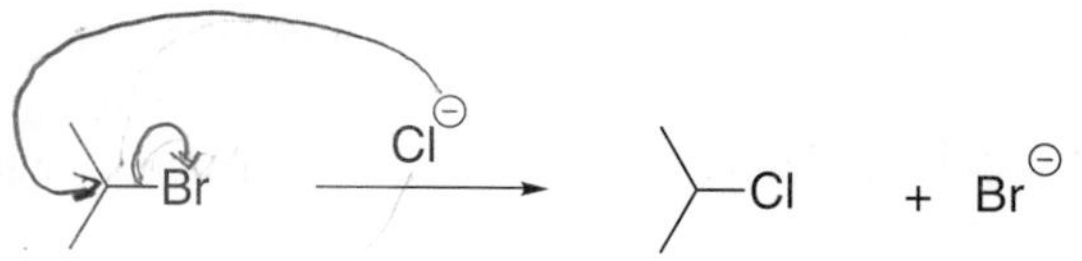

8.10

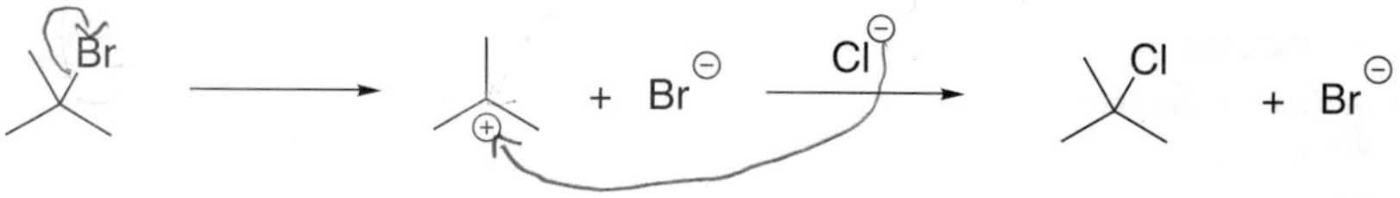

8.11

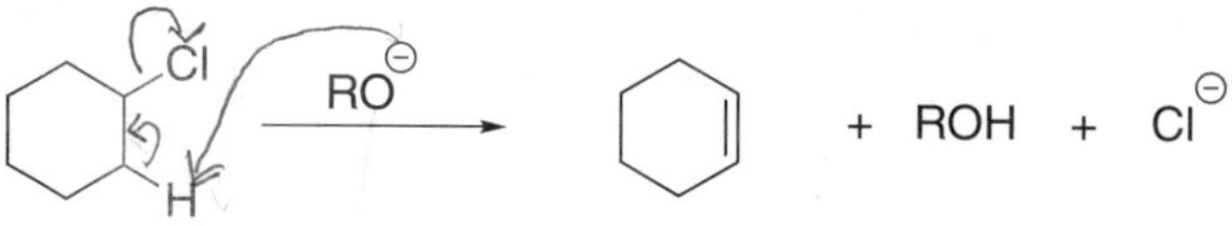

8.12

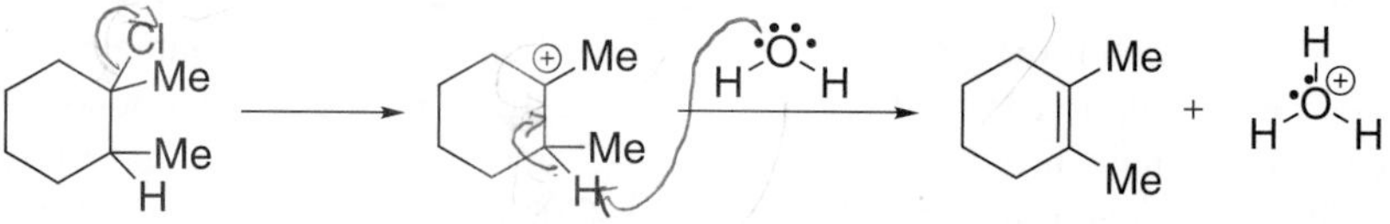

Consider the second step of problem 8.12. A lone pair from the oxygen is pulling off a proton to form a double bond:

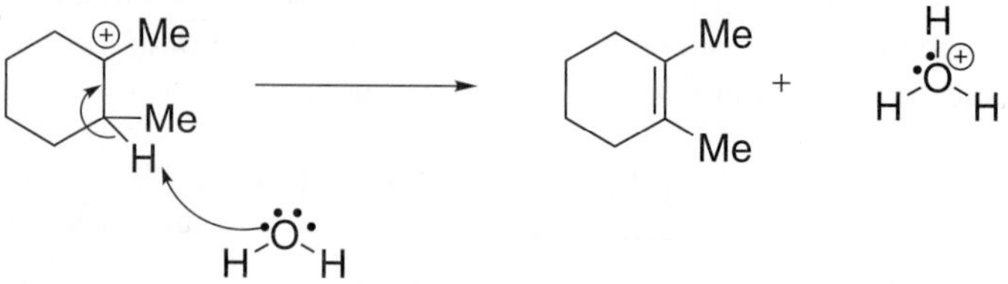

Remember that arrows indicate the flow of electrons. Arrows do not show where atoms went. Many students will accidentally draw it like this:

Students often make this mistake because they want to show where the H is going. But this is wrong. Remember that arrows show the *movement of electrons, not atoms.* The H was able to move only because the electrons came from the oxygen and grabbed the H.

8.3   DRAWING INTERMEDIATES

We have seen the different types of arrows and how to draw them. Now we need to get practice drawing intermediates when we are given the arrows. Intermediates are compounds that exist for a very short time before reacting further. Let's consider an analogy. Imagine that you are trying to climb a mountain and it is very cold (below freezing). You are wearing a hat that keeps your ears warm, but it is loose and keeps slipping off. Your friend offers you a spare hat that he brought, and you borrow it. Now you need to take your old hat off to replace it with the new hat. If someone were to take a picture of you while you have nothing on your head, the picture would look very strange. There you are, in the freezing cold, with no hat on. You were only like that for 3 seconds, but it was long enough for someone to take a picture. Intermediates of reactions are similar.

Intermediates are intermediate structures in going from the starting material to the product. They do not live for very long, and it is rare that you can isolate one and store it in a bottle, but they do exist for very short periods of time. Their structures are often critical in understanding the next step of the reaction. Going back to the analogy, if I saw the picture of you without your hat on, and I knew how cold it was on that mountain, then I would have been able to predict that you put on a hat right after the picture was taken. I would have known this because I would have been able to immediately identify an uncomfortable situation, and I could have predicted what resolution must have taken place to alleviate the problem. The same is true of intermediates. If we can look at an intermediate and determine which part of the intermediate is unstable, and we also know what options are available to alleviate the instability, then we can predict the products of the reaction based on an analysis of the intermediate. That's why they are so important.

So let's get practice drawing intermediates. If you look closely at any step of a mechanism, you will see that the arrows tell you exactly how to draw the intermediate. Since you know how to classify every arrow into one of three categories (previous section of this chapter), now you will be able to read each arrow as a road map of how to draw the intermediate. Here's an example:

Let's read the arrows. The first arrow is from a lone pair to form a bond. The arrow shows electrons in a lone pair on a nucleophile (anything that is electron rich) forming a bond with a carbon atom. The second arrow is from a bond to a bond. The third arrow goes from a bond to form a lone pair. All in all, these arrows serve as a road map for drawing the intermediate:

The trickiest part is getting the formal charges correct. If you have trouble assigning formal charges, then you will need to go back and review the sections on formal charges in Chapter 1 and Chapter 2 of this book. Assigning formal charges is a very important part of drawing the intermediates. Drawing the structure without the charges would be like taking the photograph in the analogy above, but digitally removing all of the snow. Without the snow, I wouldn't know that it was cold, so I would not be able to predict that you put a hat on shortly after the picture was taken. If you don't draw the source of instability on the intermediate, then what good is it?

One trick will help you in some situations when you have a flow of electrons represented by a few arrows (as in the example above). Notice that the only change in formal charges comes on the first and last atom of the system where the electrons are flowing. In our example above, the nucleophile loses its negative charge by using its lone pair to form a bond with a carbon atom. At the other end of the system, the oxygen is gaining a negative charge as a bond is converted into a lone pair on oxygen. Notice the conservation of charge. If the overall charge is negative at the beginning of the reaction, then it must also be negative at the end of the reaction. If something starts off with no charge, then it can split up into a positive charge and a negative charge, because the *total* charge is still conserved.

EXERCISE 8.13 Look at the arrows below, and draw the intermediate that you get after pushing the arrows:

Answer   We need to read the arrows like a road map: the first arrow is going from a lone pair on HO⁻ to form a bond with the carbon of the C=O. The second arrow goes from the C=O bond to form a lone pair on oxygen. We use this info to draw the products:

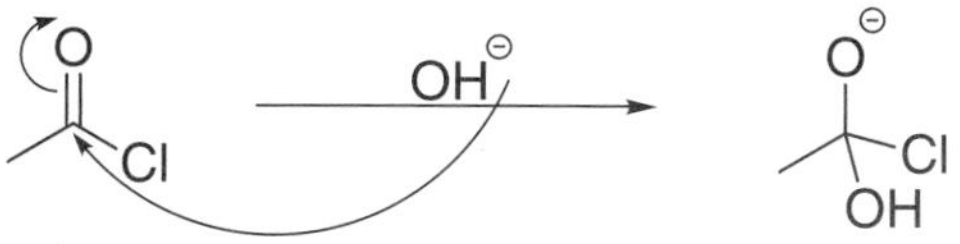

The hard part was assigning formal charges. Notice that we had two arrows moving in a flow. We had a negative charge in the beginning, so we must have a negative charge in the end. It started off on the first atom in the flow of arrows, and it ended on the last atom of the flow (the oxygen).

PROBLEMS   For each problem below, draw the intermediate that you get after pushing the arrows.

8.14

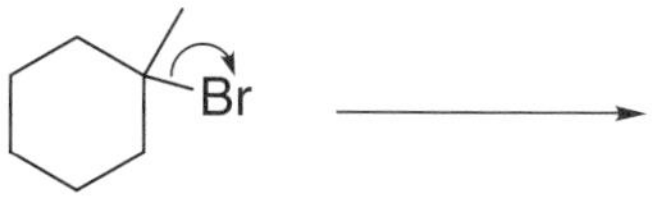

8.15

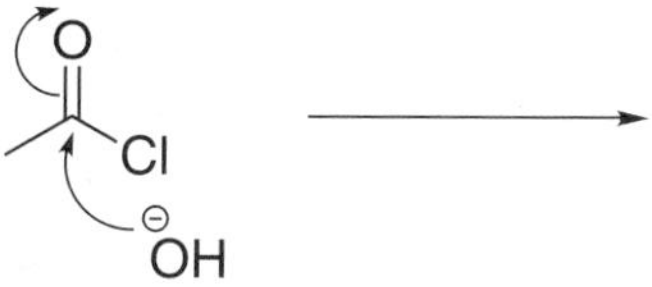

8.16

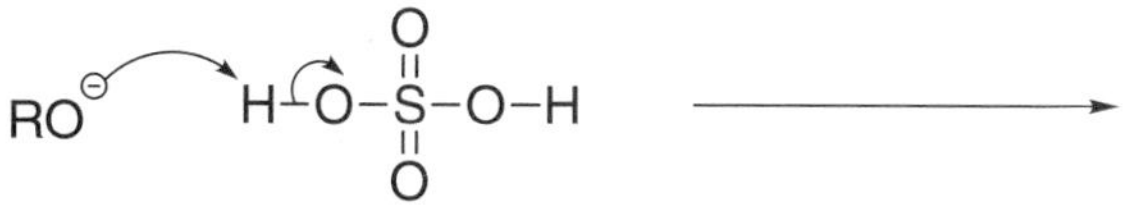

8.17

8.18

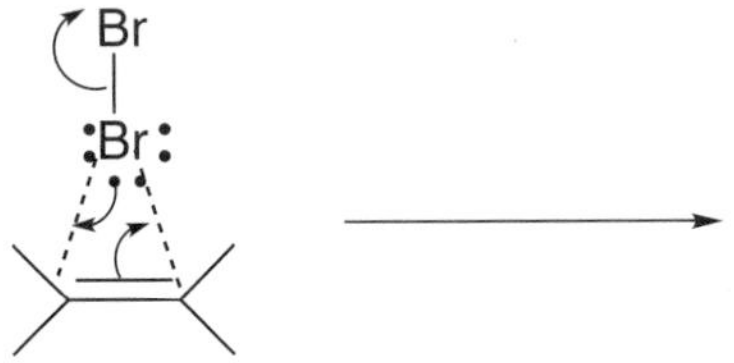

8.19

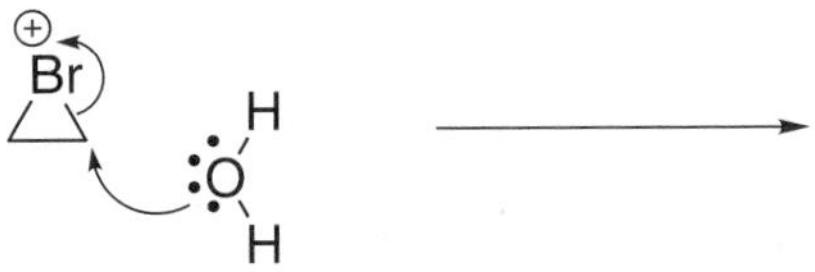

8.4 NUCLEOPHILES AND ELECTROPHILES

Whenever one compound uses its electrons to attack another compound, we call the attacker a *nucleophile,* and we call the compound being attacked an *electrophile.* It is very simple to tell the difference between an electrophile and a nucleophile. You just look at the arrows and see which compound is attacking the other. A nucleophile will always use a region of high electron density (either a lone pair or a bond) to attack the electrophile (which, by definition, has a region of low electron density that can be attacked). These are important terms, so let's make sure we know how to identify nucleophiles and electrophiles.

EXERCISE 8.20 In the reaction below, determine which compound is the nucleophile and which compound is the electrophile:

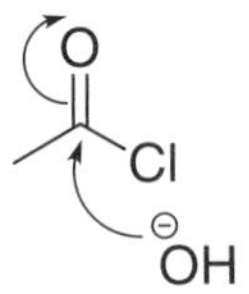

Answer The OH is attacking the C=O bond, so the OH is the nucleophile and the other compound is the electrophile:

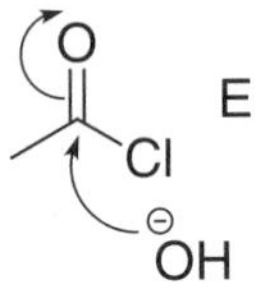

PROBLEMS   In each of the reactions below, determine which compound is the nucleophile and which compound is the electrophile.

8.21

8.22

8.23

8.24

8.5 BASES VERSUS NUCLEOPHILES

Students are often unclear about the difference between nucleophiles and bases. Since most mechanisms involve the use of nucleophiles and bases, it will be worth our time to clear up the difference.

Consider the hydroxide ion (OH^-). Sometimes it acts like a base and pulls off a proton from another compound:

$$HO^{\ominus} \quad \longrightarrow \quad H_2O \quad + \quad \text{(alkene)} \quad + \quad X^{\ominus}$$

At other times it acts like a nucleophile and attacks another compound (forming a new bond to an atom in that compound):

$$HO^{\ominus} \quad \longrightarrow \quad HO- \quad + \quad Cl^{\ominus}$$

The difference between basicity and nucleophilicity is a difference of *function*. In other words, the hydroxide ion can function in two ways: as a base (which means it is pulling off a proton and then running away with that proton) or as a nucleophile (latching onto a compound). In some cases, the hydroxide ion might function mostly as a base; while in other situations, the hydroxide ion might function mostly as a nucleophile. To understand mechanisms well, it is important to be able to distinguish between the two roles. Let's see an example.

EXERCISE 8.25 Below you will find the first two steps of a mechanism. In each step, determine whether the hydroxide ion is functioning as a nucleophile or as a base:

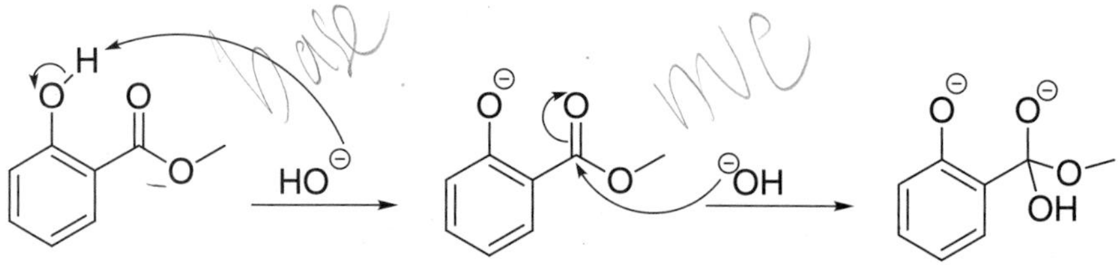

Answer In the first step, the hydroxide ion is pulling off a proton, so it is functioning as a base. In the second step, it is attacking the $C=O$ and latching on to the compound, so it is functioning as a nucleophile.

PROBLEMS In each step below, determine whether the hydroxide ion is functioning as a nucleophile or as a base.

8.26 Answer: _nuc_

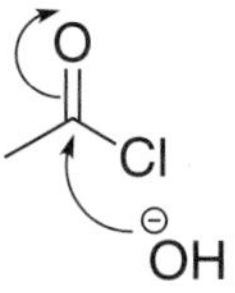

8.27 Answer: _base_

8.28 Answer: _nuc_

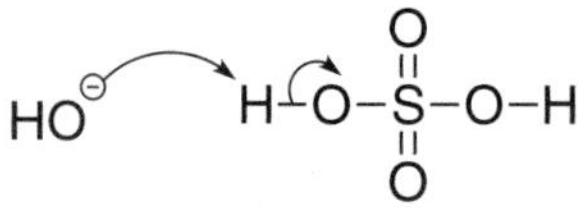

8.29 Answer: _base_

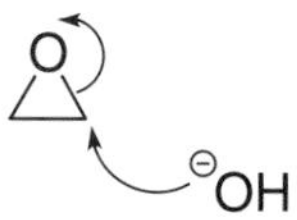

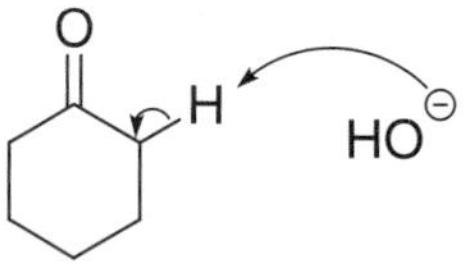

PROBLEMS In each step below, determine whether the methoxide ion (MeO⁻) is functioning as a nucleophile or as a base.

8.30 MeO Answer: _base_

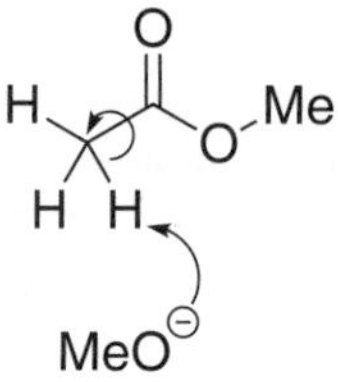

8.31

Answer: _nuc_

PROBLEMS In each step below, determine whether water is functioning as a nucleophile or as a base.

8.32

Answer: _nuc_

8.33

Answer: _base_

There is another subtle difference between nucleophiles and bases that is worth mentioning, because it illustrates a common theme in organic chemistry. We can see the difference by defining the terms nucleophilicity and basicity.

Once we determine that a reagent is acting as a nucleophile, we measure how fast it functions that way with the term *nucleophilicity*. Nucleophilicity measures *how quickly* a reagent will attack another compound. For example, we saw above that water can function as a nucleophile because it has lone pairs that can attack a compound. But the hydroxide ion will clearly be more nucleophilic—the hydroxide ion has a negative charge, so it will attack compounds *faster*.

Basicity measures base strength (or how unstable the base is) by *the position of equilibrium*. The term *basicity* does not reflect how quickly the equilibrium was reached. The equilibrium might have been established in a fraction of a second or it could have taken several hours. It doesn't matter, because we are not measuring speed of reaction. We are measuring stability and the position of the equilibrium.

Now we can understand this difference between nucleophilicity and basicity. Nucleophilicity measures how fast things happen, which is called *kinetics*. Basicity measures stability and the position of equilibrium, which is called *thermodynamics*. Throughout your course, you will see many reactions where the product is determined by kinetic concepts, and you will also see many reactions where the product is determined by thermodynamic concepts. In fact, there will even be times, where these two factors are competing with each other and you will need to make a choice of which factor wins: kinetics or thermodynamics.

So the difference between nucleophiles and bases is a difference of function. And now we can also appreciate that nucleophilicity is a measure of a kinetic phenomenon (rate of reaction), while basicity is a measure of stability (thermodynamic phenomenon).

8.6 THE REGIOCHEMISTRY IS CONTAINED WITHIN THE MECHANISM

Regiochemistry refers to *where* the reaction takes place. In other words, in what region of the molecule is the reaction taking place? Let's see examples of this for different types of reactions. In the process, we will uncover some new terminology as we learn about different reactions.

Let's consider elimination reactions. When we eliminate H and X (where X is some leaving group that can leave with a negative charge, like Cl or Br), it is possible to form the double bond in different locations. Consider the following compound:

This compound can undergo two possible elimination reactions (to make it easier to see, we are drawing the H that gets eliminated in each case, even though we usually do not draw hydrogen atoms on bond-line drawings):

Where does the double bond form? This is a question of regiochemistry. We distinguish between these two possibilities by considering how many groups are attached to each double bond. Double bonds can have anywhere from 1 to 4 groups attached to them:

| Monosubstituted | Disubstituted | Trisubstituted | Tetrasubstituted |

So if we look back at the reaction above, we find that the two possible products are monosubstituted and disubstituted double bonds. Whenever we have an elimination reaction where more than one possible double bond can be formed, we have names for the different products based on which one is more substituted and which one is less substituted. The more substituted product is called the Zaitsev product, and the less substituted product is called the Hoffmann product. Usually, we get the Zaitsev product, but under special circumstances we get the Hoffman product. You will learn about this in detail in your textbook when you cover elimination reactions. For now, you just need to realize that this is an issue of regiochemistry. The difference between the Zaitsev product and the Hoffman has to do with where the double bond formed. This is regiochemistry.

Let's consider another example of regiochemistry, in a completely different type of reaction. Consider the addition reaction of HCl across a double bond:

Cl is on the less substituted carbon

Cl is on the more substituted carbon

There are two possible ways to add the H and the Cl. Which product do we get?

One possibility would be to put the Cl on the less substituted carbon (carbon connected to two other carbon atoms), and the other possibility would be to put the Cl on the more substituted carbon (carbon connected to three other carbon atoms). If we put the Cl on the more substituted carbon, we call this a Markovnikov addition. If we put the Cl on the less substituted carbon, we call it an anti-Markovnikov addition. How do we know whether we get Markovnikov addition or anti-Markovnikov addition? This is an issue of regiochemistry.

For the reaction above, let's analyze the two possible outcomes. In each case, the first step involves the electrons of the double bond attacking the proton of HCl to form a carbocation (a carbon with a positive charge). The difference between the two possibilities is where the carbocation is formed:

Less stable

More stable

Recall that alkyl groups are electron donating, so the carbocation on the bottom (called a tertiary carbocation because it has three alkyl groups) will be more stable than the carbocation on the top (called a secondary carbocation because it has only two alkyl groups).

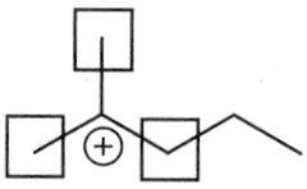

Tertiary
(more stable)

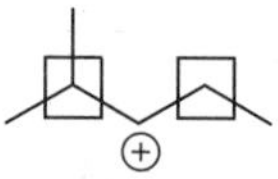

Secondary
(less stable)

Therefore, possibility 2 is a better mechanism (because it involves a more stable intermediate. If we follow the last step of the mechanism for possibility 2, we see that the Cl will attach where the carbocation is, which will be at the more substituted carbon:

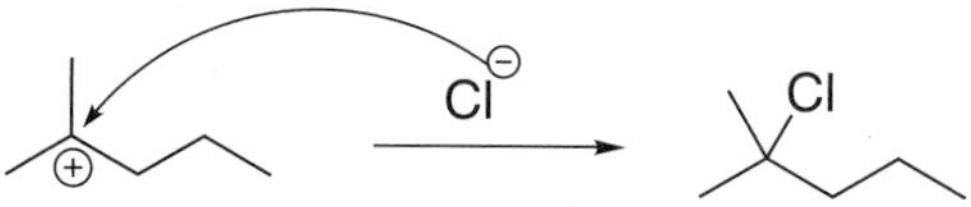

We see that the final position of the chlorine is determined by the stability of the intermediate carbocation, which becomes evident as we work through the mechanism. Since the chlorine ends up at the more substituted carbon, we call this a Markovnikov addition. The mechanism for this reaction helped explain the regiochemistry of the reaction.

Sometimes regiochemistry is not an issue. For example, if we are adding H and H across a double bond, then it does not matter which carbon gets the first H and which carbon gets the second H. Either way, they both end up with an H. Similarly, if we add two OH groups across a double bond, regiochemistry also does not matter. Any time we add two of the same group across a double bond, we do not have to worry about the regiochemistry.

Here is where we get back to mechanisms. Whether we are talking about Zaitsev vs. Hoffman elimination reactions or about Markovnikov vs. anti-Markovnikov addition reactions, the explanation of the regiochemistry for every reaction is contained within the mechanism. If we completely understand the mechanism, then we will understand why the regiochemistry had to be the way it turned out. By understanding the mechanism, we eliminate the need to memorize the regiochemistry for every reaction. With every reaction you encounter, you should consider the regiochemistry of the reaction and look at the mechanism for an explanation of the regiochemistry.

PROBLEMS   You will, over the course of your studies, learn the mechanisms for the following reactions. In the meantime, you will be given the regiochemical information that you need to answer each of the problems below. These problems are intended to ensure that you understand what regiochemistry means.

8.34 Consider the reaction shown. If you were to add HBr across the double bond, what would the product be? Assume a Markovnikov addition.

H—Br

8.35 When you do the same reaction (as above) in the presence of peroxides (R-O-O-R), you get an anti-Markovnikov addition of HBr across the double bond. Draw the product of an anti-Markovnikov addition.

H—Br

ROOR

8.36 Consider the elimination reaction below, which uses a strong base. The product will be a double bond. This reaction will produce two Zaitsev products. One will be cis and one will be trans. Draw these products, and identify which is cis and which is trans.

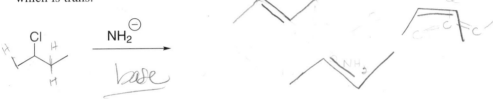

Cl

NH₂

8.37 Consider the elimination reaction below, which uses a strong, sterically hindered base (LDA). The product will be a double bond. This reaction will produce the Hoffmann product. Draw this product.

Cl

LDA

8.7 THE STEREOCHEMISTRY IS CONTAINED WITHIN THE MECHANISM

Stereochemistry is all about configurations of stereocenters (R vs. S) and double bonds (E vs. Z). Whenever we have a reaction where we are forming a stereocenter, we need to ask whether we get a racemic mixture (equal amounts of R and S) or only one configuration. And, if so, why? Also, whenever we form a double bond, we need to ask whether we get both E and Z isomers or only one of them? And, if so, why?

This information is also contained within the mechanism. Let's see an example. Consider the addition of Br and Br across a double bond. We already saw that we don't need to worry about the regiochemistry of this reaction, because we are adding two of the same group. But what about the stereochemistry? We are creating two new stereocenters:

Each stereocenter has two possibilities (R or S). Since there are two stereocenters, we will have four total possibilities: SR, RS, RR, and SS. These four compounds represent two sets of enantiomers:

One set of enantiomers,	Another set of enantiomers,
Br and Br are on the same side of the ring	Br and Br are on opposite sides of the ring

How many of them do we get? Do we get both sets of enantiomers as our products (meaning all four products), or do we only get one set (meaning two out of the four possible products)? This depends on how the reaction took place.

If an addition reaction can take place only through a mechanism that allows a syn addition, then the two groups that we added must be on the same side of the double bond in the product. So we will get only that one set of enantiomers:

If a reaction can go through only an anti addition, then the two groups we added must be on opposite sides of the double bond. So we will get only the set of enantiomers where the groups are on opposite sides:

Sometimes, the reaction is not stereoselective. In other words, we get both syn and anti addition. So we get all four products (both sets of enantiomers).

Each reaction will be different. Some will give only syn addition, some will give only anti addition, and others will not be stereoselective. For every addition reaction, we need to know the stereochemistry of the addition, and that information is contained within the mechanism.

So let's go back to our example above with the addition of Br and Br across a double bond. This reaction is an anti addition, so we get only the set of enantiomers that has the two Br groups on opposite sides of the ring:

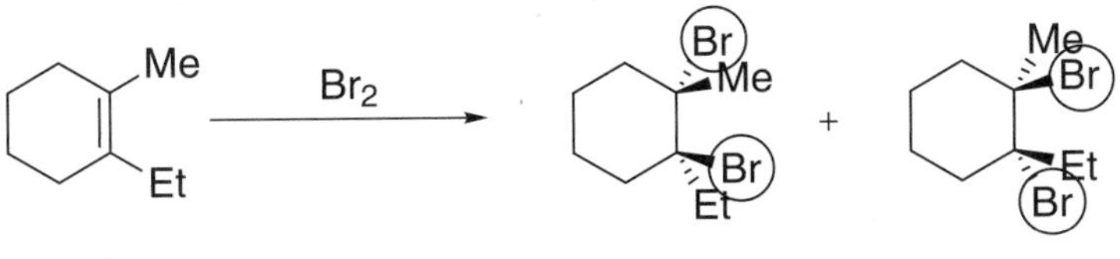

Let's look at the mechanism to understand why. In the first step, we form a bridged intermediate, called a bromonium ion:

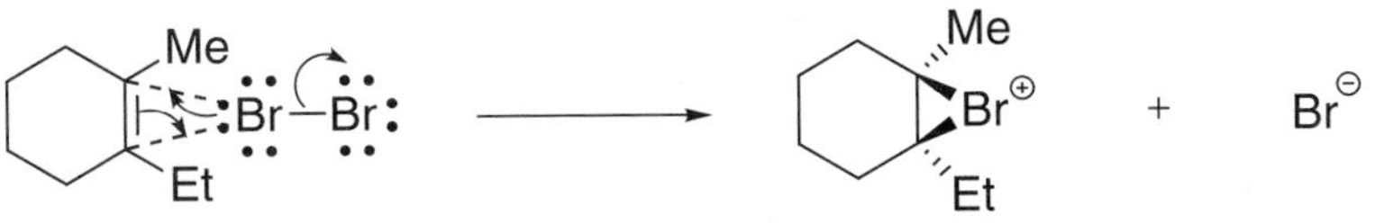

In this step, the double bond is acting as a nucleophile that attacks Br_2 (the electrophile in this reaction). The arrows are not going in opposite directions—they are actually moving in a small circle to form a ring.

Then, in the next step, the bromide (formed in the first step) comes back and attacks the bromonium ion, opening up the bridge. The bromide can attack either carbon (both possibilities shown below):

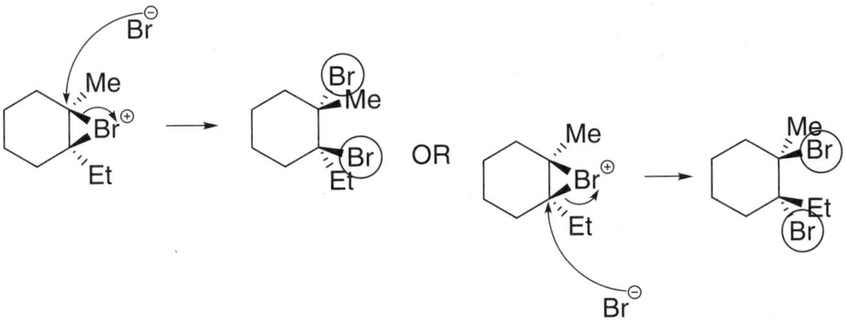

When the bromide attacks, it must attack on the other side of the ring (not the side of the bromonium bridge, but rather on the opposite side of the ring) to break open the bridge. So the addition must be an *anti addition*.

We see that the mechanism explains why the addition must be anti. For every reaction, the stereochemistry will always be explained by the mechanism.

PROBLEM 8.38 In the reaction above, we saw that the first step involved formation of a bromonium ion.

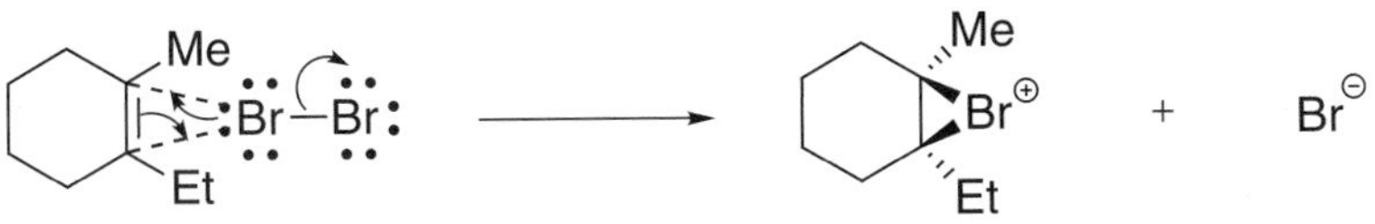

You will notice that the bromonium ion drawn above has the bridge coming out towards you (on wedges), but we did not say at the time that it could also have formed with the bridge going away from you (on dashes):

We did not talk about this at the time, because the end products would still have been the same as the way we did it before. Draw what happens if the bromide (Br⁻) attacks this other bromonium ion. Remember that there are two carbon atoms that the bromide could attack, so draw both possibilities:

When you finish drawing the two products, compare them to the two products that we got before. You should find that the two products you get here are the same as the two products we got before. Think about why. Remember that the reaction can happen only as an anti addition.

Every new class of reactions (additions, eliminations, substitutions, etc.) has its own terminology for stereochemistry. As you learn each of these classes of reactions, keep a watchful eye on what terminology is used to describe the stereochemistry. Then, look at the mechanism of each reaction within each class, and try to understand why the mechanism dictates the stereochemistry.

PROBLEMS   For the following reactions, you will (over the course of your studies) learn the mechanisms for these reactions. In the meantime, you will be given the stereochemical information that you need to answer each of the problems below. These problems are intended to ensure that you understand what stereochemistry means.

8.39 If you add OH and OH across the following double bond in a syn addition, what will the products be?

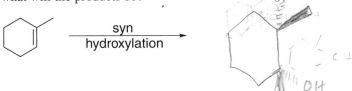

syn
hydroxylation

8.40 If you add Br and Br across the following double bond in an anti addition, what will the products be?

Me   Me
        Br_2
Et   Et

8.41 If you add Br and Br across the following double bond in an anti addition, you get only one product. If you draw the two products that you would expect, you will find that they are the same compound (a meso compound). Draw this product.

Me   Et
        Br_2
Et   Me

Do not confuse the concepts of regiochemistry and stereochemistry. For instance, in addition reactions, the term "anti-Markovnikov addition" refers to the *regiochemistry* of the addition, but the term "anti" refers to the *stereochemistry* of the addition. Students often confuse these concepts (probably because both terms have the word "anti"). It is possible for an addition reaction to be anti-Markovnikov and a syn addition (hydroboration is an example that you will learn about at some point in time). You must realize that regiochemistry and stereochemistry are two totally different concepts.

8.42 In the following reaction, we will add H and OH across a double bond. The regiochemistry is anti-Markovnikov, and the stereochemistry is a syn addition. Draw the products you would expect now that you know all of the information.

(1) BH_3 / THF
(2) H_2O_2 / OH^-

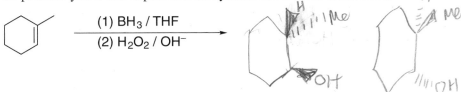

You must know the stereochemistry and regiochemistry for every reaction, and each of them is contained within the mechanism. In the problems above, you were told what to expect for the stereochemistry and the regiochemistry. When you are doing problems in your textbook and on your exams, you will be expected to know what these pieces of information are simply from looking at the reagents. A solid understanding of every mechanism will be an invaluable asset to you in this course.

8.8 A LIST OF MECHANISMS

Now you need to begin to keep a list of all reaction mechanisms that you cover. The rest of the pages in this chapter are set up specifically for you to generate this list in such a way that you will record the critical information: the regiochemistry and the stereochemistry. You should fill in these pages as you proceed through the course and you learn more mechanisms. As your list gets larger, you will have one central place where you can go to review all of the mechanisms.

A few example mechanisms have been filled in for you, so that you can see how to fill in the each mechanism from now on. Depending on the order that your course follows, these reactions may or may not be the first ones you cover. Whatever the case might be, you will definitely see these reactions early on in the course:

Reaction type	Stereochemistry	Regiochemistry
Substitution (S_N2)	Inversion	Not applicable (nucleophile attacks carbon next to LG)

Reaction type	Stereochemistry	Regiochemistry
Substitution (S_N1)	Racemization	Not applicable (nucleophile attacks carbon next to LG)

Now, for every reaction that you cover, fill in the templates below, and then use this list as a study guide for all of your mechanisms:

Reaction type	Stereochemistry	Regiochemistry

Reaction type	Stereochemistry	Regiochemistry

Reaction type	Stereochemistry	Regiochemistry

Reaction type	Stereochemistry	Regiochemistry

Reaction type	Stereochemistry	Regiochemistry

Reaction type	Stereochemistry	Regiochemistry

Reaction type	Stereochemistry	Regiochemistry

Reaction type	Stereochemistry	Regiochemistry

Reaction type	Stereochemistry	Regiochemistry

Reaction type	Stereochemistry	Regiochemistry

Reaction type	Stereochemistry	Regiochemistry

Reaction type	Stereochemistry	Regiochemistry

Reaction type	Stereochemistry	Regiochemistry

Reaction type	Stereochemistry	Regiochemistry

Reaction type	Stereochemistry	Regiochemistry

Reaction type	Stereochemistry	Regiochemistry

Reaction type	Stereochemistry	Regiochemistry

Reaction type	Stereochemistry	Regiochemistry

Reaction type	Stereochemistry	Regiochemistry

Reaction type	Stereochemistry	Regiochemistry

Reaction type	Stereochemistry	Regiochemistry

Reaction type	Stereochemistry	Regiochemistry

Reaction type	Stereochemistry	Regiochemistry

Reaction type	Stereochemistry	Regiochemistry

Reaction type	Stereochemistry	Regiochemistry

Reaction type	Stereochemistry	Regiochemistry

Reaction type	Stereochemistry	Regiochemistry

Reaction type	Stereochemistry	Regiochemistry

Reaction type	Stereochemistry	Regiochemistry

Reaction type	Stereochemistry	Regiochemistry

Reaction type	Stereochemistry	Regiochemistry

Reaction type	Stereochemistry	Regiochemistry

Reaction type	Stereochemistry	Regiochemistry

Reaction type	Stereochemistry	Regiochemistry

Reaction type	Stereochemistry	Regiochemistry

Reaction type	Stereochemistry	Regiochemistry

Reaction type	Stereochemistry	Regiochemistry

Reaction type	Stereochemistry	Regiochemistry

Reaction type	Stereochemistry	Regiochemistry

Reaction type	Stereochemistry	Regiochemistry

CHAPTER 9

SUBSTITUTION REACTIONS

In the last chapter we saw the importance of understanding mechanisms. We said that mechanisms are the keys to understanding everything else. In this chapter, we will see a very special case of this. Students often have difficulty with substitution reactions—specifically, being able to predict whether a reaction is an S_N2 or an S_N1. These are different types of substitution reactions and their mechanisms are very different from each other. By focusing on the differences in their mechanisms, we can understand why we get S_N2 in some cases and S_N1 in other cases.

Four factors are used to determine which reaction takes place. These four factors make perfect sense when we understand the mechanisms. So, it makes sense to start off with the mechanisms.

9.1 THE MECHANISMS

Ninety-five percent of the reactions that we see in organic chemistry occur between a nucleophile and an electrophile. A nucleophile is a compound that either is negatively charged or has a region of high electron density (like a lone pair or a double bond). An electrophile is a compound that either is positively charged or has a region of low electron density. When a nucleophile and an electrophile find each other space, they can attract each other (opposite charges). If the conditions are right, we can have a reaction between them.

In both S_N2 and S_N1 reactions, a *nucleophile* is attacking an electrophile, giving us a *substitution* reaction. That explains the S_N part of the name. But what do the "1" and "2" stand for? To see this, we need to look at the mechanisms. Let's start with S_N2:

On the left, we see a nucleophile. It is attacking a compound that has an electrophilic carbon that is attached to a leaving group (LG). A *leaving group* is any group that is willing to be kicked off (we will see examples of this very soon). This leaving group is generally electronegative (because it needs to be happy leaving with a negative charge), which is why the carbon is electrophilic. The leaving group is withdrawing electron density from the carbon, making it electron poor.

Our S_N2 mechanism has two arrows: one going from a lone pair on the nucleophile to form a bond between the nucleophile and carbon, and the other going from the bond between the carbon and the LG to form a lone pair on the LG. Notice that the configuration at the carbon atom gets inverted in this reaction. So the stereochemistry of this reaction is inversion of configuration. Why does this happen? It is kind of like an umbrella flipping inside out in a strong wind. It takes a good force to do it, but it is possible to flip the umbrella. The same is true here. If the nucleophile is good enough, and if all of the other conditions are just right, we can invert the stereocenter (by bringing the nucleophile in on one side, and kicking off the LG on the other side).

Now we get to the meaning of "2" in S_N2. Remember from the last chapter that nucleophilicity is a measure of kinetics (how fast something happens). Since this is a *nucleophilic* substitution reaction, then we care about how fast the reaction is happening. In other words, what is the rate of the reaction? This mechanism has only one step, and in that step, two things need to find each other: the nucleophile and the electrophile. So it makes sense that the rate of the reaction will be dependent on how much electrophile is around *and* how much nucleophile is around. In other words, the rate of the reaction is dependent on the concentrations of two compounds. Therefore, we call the reaction bimolecular and we put a "2" in the name of the reaction.

Now let's look at the mechanism for an S_N1 reaction:

In this reaction, there are two steps. The first step has the LG leaving all by itself, without any help from an attacking nucleophile. This leaves behind a carbocation, which then gets attacked by the nucleophile in step 2. This is the major difference between S_N2 and S_N1 reactions. In S_N2 reactions, everything happens in one step. In S_N1 reactions, it happens in two steps, and we are forming a carbocation in the process. The existence of the carbocation as an intermediate in *only* the S_N1 mechanism is the key. By understanding this, we can understand everything else.

For example, let's look at the stereochemistry of the S_N1. We already saw that the S_N2 reaction went through inversion of configuration. But the S_N1 reaction is very different. Recall that a carbocation is sp^2 hybridized, so its geometry is trigonal planar. When the nucleophile attacks, there is no preference as to which side it can attack, and we get both possible configurations in equal amounts. Half of the molecules would have one configuration and the other half would have the other configuration. We learned before that this is called a racemic mixture. Notice that we can explain the stereochemical outcome of this reaction by understanding the nature of the carbocation intermediate that is formed.

This also allows us to understand why we have the "1" in S_N1. There are two steps in this reaction. The first step is very slow (the LG just leaves on its own to form C^+ and LG^-, which is pretty strange when you think about it), and the second step is very fast. Therefore, the rate of the second step is irrelevant. Let's use an analogy to understand this.

Imagine that you have an hourglass with two openings that the sand had to pass by:

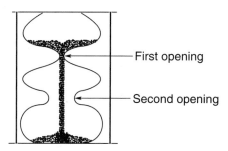

The first opening is much smaller, and the sand can travel through this opening only at a certain speed. The size of the second opening doesn't really matter. If you made the second opening a little bit wider, it would not help the sand get to the bottom any faster. As long as the top opening is smaller, the rate of the falling sand will depend only on the size of the top opening.

The same is true in a two-step reaction. If the first step is slow and the second step is fast, then the speed of the second step is irrelevant. The rate at which you get product will depend *only* on the rate of the first step (the slow step). So in our S_N1 reaction, the first step is the slow step (loss of the LG to form the carbocation) and the second step is fast (nucleophile attacking the carbocation). Just as we saw in the hourglass, the second step of our mechanism will not affect the rate of the reaction. Notice that the nucleophile does not appear in the mechanism until the second step. If we added more nucleophile, it would not affect the rate of the first step. Adding more nucleophile would only speed up the second step. But we already saw that the rate of the second step does not matter for the overall reaction rate. Speeding up the second step will not change anything. So the concentration of nucleophile does not affect the rate of the reaction.

Of course, it is important that we have a nucleophile present, but how much we have doesn't matter. So now we can understand the "1" in S_N1. The *rate* of the reaction is dependent only on the concentration of the electrophile, and not that of the nucleophile. Since the rate is dependent on the concentration of only one thing, we call the reaction unimolecular, and we put a "1" in the name. Of course, this does not mean that you *only* need the electrophile. You still need the nucleophile for the reaction to happen. You still need two different things (nucleophile and electrophile). The "1" simply means that the rate is not dependent on the concentration of both of them. The rate is dependent on the concentration of only one of them.

The mechanisms of the S_N1 and S_N2 reaction helped us understand the stereochemistry of each reaction, and we were also able to see why we call them S_N1 and S_N2 reactions (based on reaction rates that are justified by looking at the mechanisms). So, the mechanisms really do explain a lot.

We mentioned before that we need to consider four factors when choosing whether a reaction will go by an S_N1 or S_N2 mechanism. These four factors are: electrophile, nucleophile, leaving group, and solvent. We will go through each factor one at a time, and we will see that the difference between the two mechanisms is the key to understanding each of these four factors. Before we move on, it is very important that you understand the two mechanisms. For practice, try to draw them in the space below without looking back to see them again.

Remember, an S_N2 mechanism has one step: the nucleophile attacks the electrophile, kicking off the leaving group. An S_N1 mechanism has two steps: first the leaving groups leaves to form a carbocation, and then the nucleophile attacks that carbocation. Also remember that S_N2 involves inversion of configuration, while S_N1 involves racemization. Now, try to draw them.

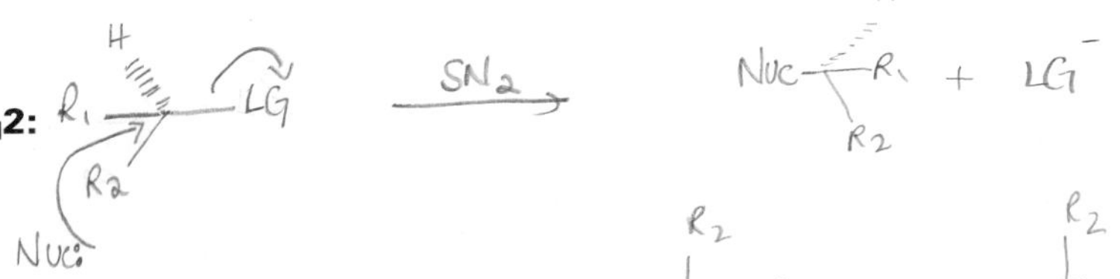

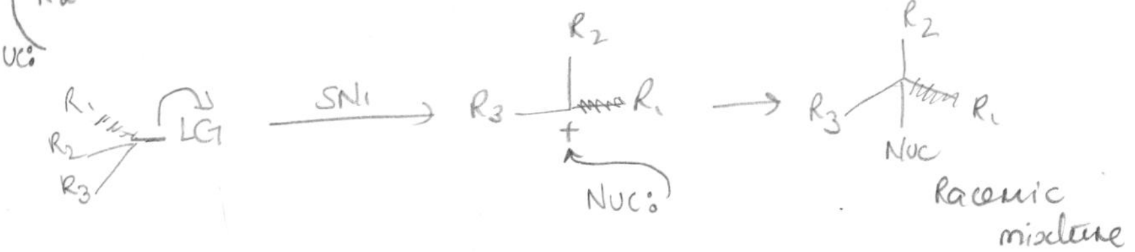

9.2   FACTOR 1—THE ELECTROPHILE (SUBSTRATE)

The electrophile is the compound being attacked by the nucleophile. In substitution and elimination reactions (which we will see in the next chapter), we generally refer to the electrophile as the *substrate*.

Remember that carbon has four bonds. So, other than the bond to the leaving group, the carbon atom that we are attacking has three other bonds:

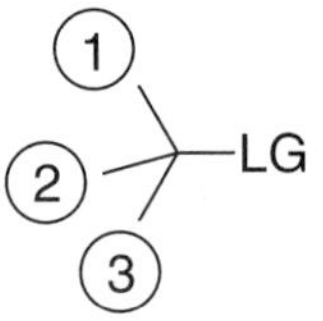

The question is, how many of these groups are alkyl groups (methyl, ethyl, propyl, etc.)? We represent alkyl groups with the letter "R". If there is one alkyl group, we call the substrate "primary" (1°). If there are two alkyl groups, we call the substrate "secondary" (2°). And if there are three alkyl groups, we call the substrate "tertiary" (3°):

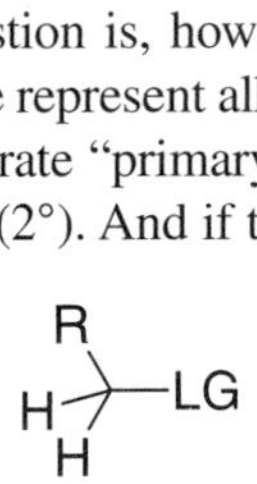

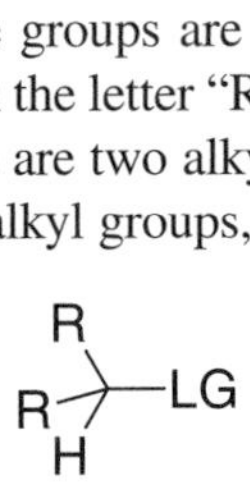

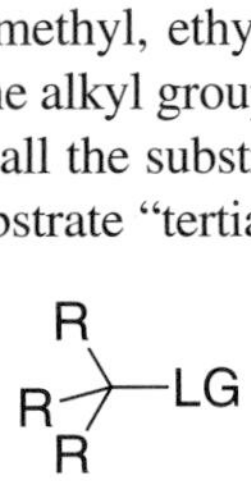

In an S_N2 reaction, alkyl groups make it very crowded at the electrophilic center where the nucleophile needs to attack. If there are three alkyl groups, then it is virtually impossible for the nucleophile to get in and attack (this is an argument based on sterics):

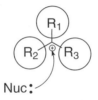

So, for S_N2 reactions, 1° is the best, 2° is OK, and 3° rarely happens.

But S_N1 reactions are totally different. The first step is not attack of the nucleophile. The first step is loss of the leaving group to form the carbocation. Then the nucleophile attacks the carbocation. Remember that carbocations are trigonal planar, so it doesn't matter how big the groups are. The groups go out into the plane, so it is easy for the nucleophile to attack. Sterics is not a problem:

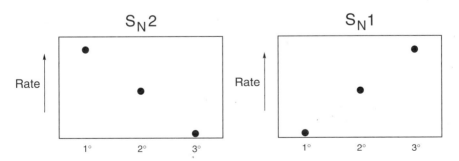

In S_N1 reactions, the stability of the carbocation is the _paramount_ issue. *chief in importance.* Recall that alkyl groups are electron donating. Therefore, 3° is best because the three alkyl groups stabilize the carbocation. 1° is the worst because there is only one alkyl group to stabilize the carbocation. This has nothing to do with sterics; this is an argument of electronics (stability of charge). So we have two opposite trends, for completely different reasons:

These charts show the rate of reaction. If you have a 1° substrate, then the reaction will go via an S_N2 mechanism, with inversion of configuration. If you have a 3° substrate, then the reaction will go via an S_N1 mechanism, with racemization. What do you do if the substrate is 2°? You move on to factor 2.

EXERCISE 9.1   For the following compound, determine whether a nucleophile is more likely to attack it in an S_N2 or an S_N1 mechanism:

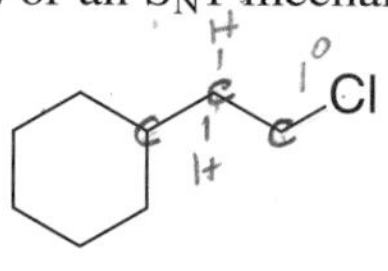

Answer   The substrate is primary, so we predict an S_N2 reaction.

PROBLEMS   For each compound below, determine whether a nucleophile is more likely to attack it in an S_N2 or an S_N1 mechanism or whether both mechanisms are possible.

9.2 _____

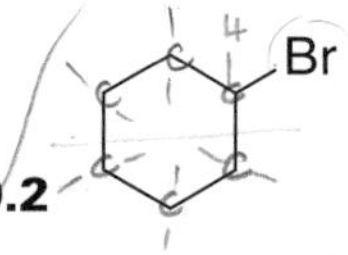

9.3 _____

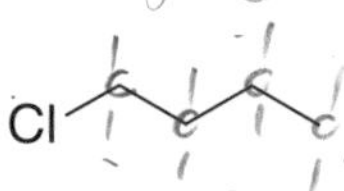

9.4 _____

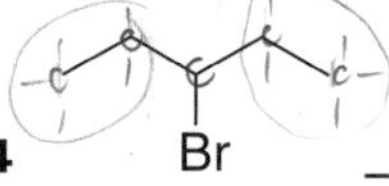

9.5 _____

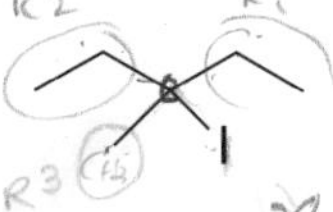

There is one other way to stabilize a carbocation (other than alkyl groups)—resonance. If a carbocation is resonance stabilized, then it will be easier to form that carbocation:

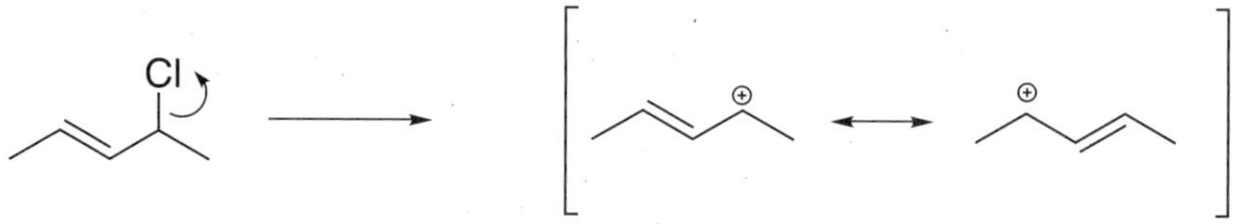

The carbocation above is stabilized by resonance. Therefore, the LG is willing to leave, and we can have an S_N1 reaction.

There are two kinds of systems that you should learn to recognize: an LG in a benzylic position and an LG in an allylic position. Compounds like this will be resonance stabilized when the LG leaves:

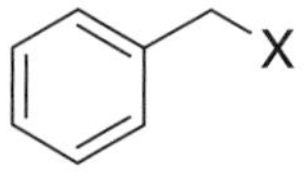

Benzylic 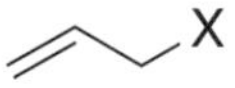   Allylic

If you see a double bond near the LG and you are not sure if it is a benzylic or allylic system, just draw the carbocation you would get and see if there are any resonance structures.

EXERCISE 9.6 In the compound below, circle the LGs that are benzylic or allylic:

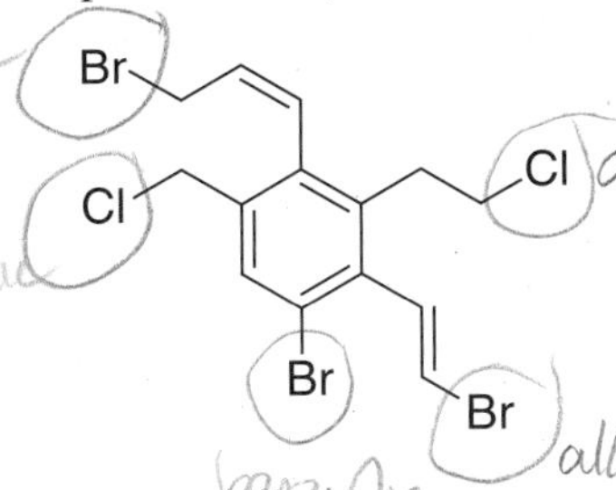

Answer

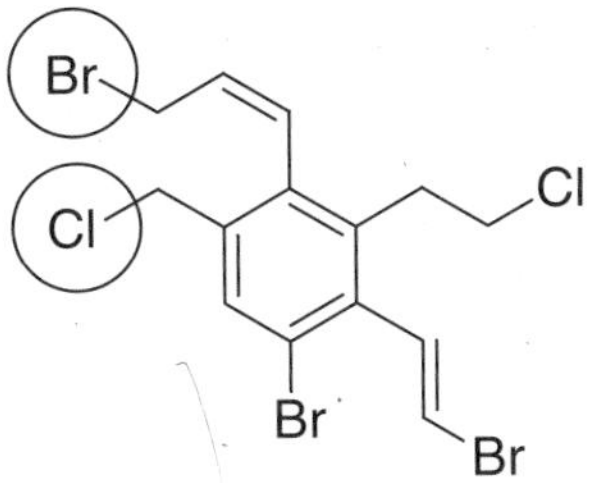

PROBLEMS For each compound below, determine whether the LG leaving would form a resonance-stabilized carbocation. If you are not sure, try to draw resonance structures of the carbocation you would get if you pulled off the leaving group.

9.7

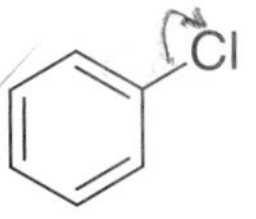

9.8

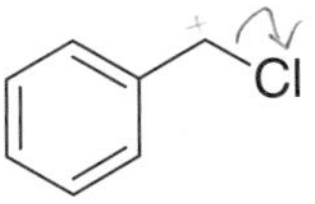

9.9 ⌇Br

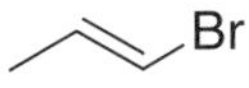

9.10 ⌇Br

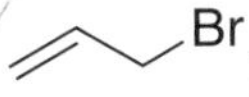

9.3 FACTOR 2—THE NUCLEOPHILE

In many cases we can determine from the substrate alone whether we will get S_N1 or S_N2. If we have a 1° substrate, then the reaction will go via an S_N2 mechanism, with inversion of configuration. If you have a 3° substrate, then the reaction will go via an S_N1 mechanism, with racemization. What do we do if the substrate is 2°? We move on to factor 2—the nucleophile. There are three categories of nucleophiles: very strong, moderate, and weak. Let's go through each of these categories, one at a time, and then we will see how the strength of the nucleophile helps us determine whether the reaction will be S_N1 or S_N2.

Weak nucleophiles are those that do not have a negative charge at all. They have lone pairs, and they use these lone pairs to attack the electrophilic site of the substrate. Examples are

Each of these compounds has no charge. Students often forget that compounds like this can be nucleophiles. But since there is no charge, they are certainly very weak nucleophiles.

Then we have the moderate nucleophiles. These are nucleophiles with a negative charge that is very stable. The halogens (F, Cl, Br, I) are excellent examples. When they have a negative charge, they are fairly stable. Other examples include resonance-stabilized ions, where the charge is spread out over more than one electronegative atom:

Finally, there are strong nucleophiles to consider. These are nucleophiles with a negative charge that is not stabilized. The charge is not on a halogen, and there is no resonance that spreads the charge out. Examples include

To put everything together, weak nucleophiles have no charge at all, moderate nucleophiles have a stabilized negative charge, and strong nucleophiles have an unstabilized negative charge. Now let's see what effect this has.

Once again, we turn to the mechanisms to understand the nucleophile's effect on the rates of reaction. We have actually already discussed this, when we explained the meaning of the "1" and the "2" in the names S_N1 and S_N2. Remember that the "2" in S_N2 meant that the rate of the reaction is dependent on the concentrations of the substrate *and* the nucleophile. If we increased the concentration of the nucleophile, we would speed up the reaction. But in an S_N1 reaction, the "1" meant that the rate of reaction is dependent only on the concentration of the substrate (remember the hourglass). The concentration of the nucleophile was not relevant in determining the rate of reaction.

Based on this, it makes sense to say that the strength of the nucleophile will make a difference *only* for S_N2. If we use a strong nucleophile, then the S_N2 rate will be fast. If we use a weak nucleophile, then the rate will be slow. But in an S_N1 reaction, it does not matter at all how strong or weak our nucleophile is. It only matters that a nucleophile is present (once the carbocation forms, it is not very picky—it will take any nucleophile it finds, whether the nucleophile has a negative charge or not). So the trends are as follows:

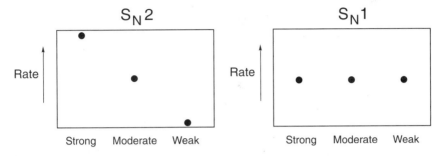

We see from these trends that a strong nucleophile favors S_N2. We often say that a weak nucleophile favors S_N1, but we really mean that a weak nucleophile *disfavors* S_N2. Therefore, if there is a competition between S_N2 and S_N1, then S_N1 will be comparatively faster than an S_N2 if we use a weak nucleophile, because the rate of an S_N2 will be very slow.

Here is the bottom line: a strong nucleophile suggests an S_N2 mechanism, and a weak nucleophile suggests an S_N1 mechanism. What do we do if the nucleophile is moderate? Move on to factor 3.

EXERCISE 9.11   Predict whether the nucleophile below will favor S_N2 or S_N1:

$$NH_2^{\ominus} \qquad SN\,2$$

Answer   There is a charge, so we know right away that this is not a weak nucleophile. Now we need to ask whether the charge is stabilized. We said that "stabilized" meant that the charge either was on a halogen or was resonance stabilized on more than one electronegative atom. In this case, the charge is not on a halogen, and the charge is not resonance stabilized. Therefore, this is a strong nucleophile. Strong nucleophiles favor S_N2 reactions.

PROBLEMS   For each nucleophile below, predict whether the nucleophile will favor S_N1 or S_N2. Identify cases where you cannot tell (and you would need to move on to factor 3).

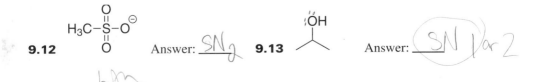

9.12 $H_3C-\overset{\overset{\displaystyle O}{\|}}{\underset{\underset{\displaystyle O}{\|}}{S}}-O^{\ominus}$ Answer: SN_2 **9.13**   Answer: $SN\ 1\,or\,2$

9.14 Answer: _SN2_   **9.15** Answer: _SN2_

9.16 Answer: _SN1_   **9.17** Br Answer: _SN1 or 2_

9.4 FACTOR 3—THE LEAVING GROUP

In many cases we can determine from the substrate and the nucleophile whether we will get S_N1 or S_N2. If we have a 1° substrate and a strong nucleophile, then the reaction will go via an S_N2 mechanism, with inversion of configuration. If we have a 3° substrate and a weak nucleophile, then the reaction will go via an S_N1 mechanism, with racemization. What do we do if the substrate is 2° and the nucleophile is moderate? We move on to factor 3—the leaving group.

There are three categories of leaving group: excellent, good, and bad. These three categories are exactly the same as the three categories of nucleophile that we just learned above. This makes sense if we think about the reactions as movies that can be played in reverse. If you play an S_N2 in reverse, the roles of nucleophile and leaving group are reversed. The same thing happens when you play an S_N1 in reverse. Go back to the two mechanisms and try to play both reactions forward and backward in your mind. You should be able to see that for S_N1 and for S_N2, the forward and reverse reactions simply exchange the roles of the nucleophile and the leaving group.

Let's go through all three categories once more. Excellent leaving groups are those that do not have a negative charge at all. Examples are

Then we have the good leaving groups. These are leaving groups with a negative charge that is very stable. The halogens (F, Cl, Br, I) are examples. When they have a negative charge, they are fairly stable. Other examples include resonance stabilized ions, where the charge is spread out over more than one electronegative atom:

Finally, there are bad leaving groups. These are nucleophiles with a negative charge that is not stabilized. The charge is not on a halogen, and there is no resonance that spreads the charge out. Examples include

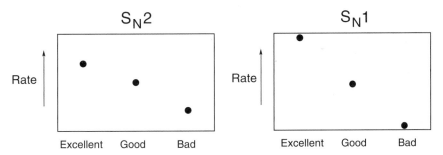

To summarize, excellent leaving groups have no charge at all, good leaving groups have a stabilized negative charge, and bad leaving groups have an unstabilized negative charge.

Now let's see what effect the leaving group has on the rates of S_N1 and S_N2. Once again, we turn to the mechanisms to understand the leaving group's effect on the rates of reaction.

The S_N2 reaction certainly needs at least a good leaving group to work, but the S_N1 reaction is much more sensitive to the stability of the leaving group. The first and only step of an S_N2 reaction is the nucleophile coming in and displacing the leaving group. As long as the leaving group is more stable than the nucleophile, then the reaction can go. But in an S_N1 reaction, the first (and rate-determining) step is loss of the leaving group. The more stable the leaving group is, the faster the reaction will happen. So, for an S_N1 reaction, the leaving group should really be excellent (although it can still happen for good leaving groups).

Let's compare the trends:

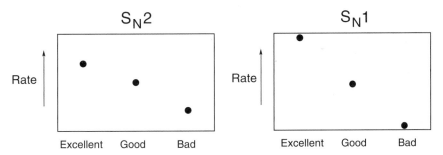

Both reactions are faster when the leaving group is better. The trends are the same, but S_N1 is more sensitive to this factor.

At this point, it is worth noting the differences in the trends for S_N2 and S_N1 when it comes to each of the factors we have seen. For the substrate (factor 1), we saw opposite trends for S_N1 and S_N2. For the nucleophile (factor 2), we saw that S_N1 did not have a trend but S_N2 did. In that case we could either favor S_N2 or disfavor S_N2. For the leaving group (factor 3), the trends for S_N2 and S_N1 are similar, but the S_N1 trend is stronger. If this sounds like jibberish to you, perhaps you should look at the charts showing these trends on the previous few pages. Compare each of the factors to each other.

Bottom line for factor 3: if you want to favor an S_N1 mechanism over an S_N2, you should really use an excellent leaving group. If you use a good leaving group, it is hard to say from this factor alone which mechanism would predominate. If you use a bad leaving group, neither reaction will occur at all.

EXERCISE 9.18 Consider the compound shown below, and predict whether the leaving group will favor S_N2 or S_N1:

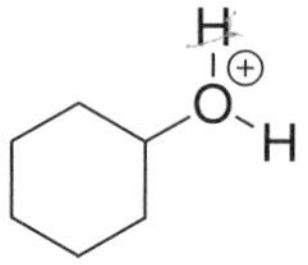

Answer Whenever you are assessing a leaving group or a nucleophile to determine how stable it is, you need to look at what it looks like when it is *not* attached to the molecule. In other words, how stable is it when it is by itself. Let's draw what we get if the leaving group leaves:

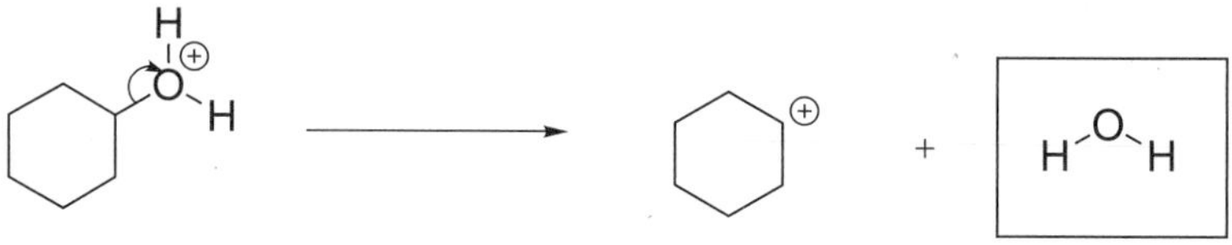

We get H_2O, which is neutral. It does not have a negative charge, so this is an excellent leaving group. For a leaving group to be neutral when it leaves, it must have a positive charge while it is still attached to the molecule. But remember that you need to look at it *after* it leaves.

An excellent leaving group favors both S_N2 and S_N1, but we said that S_N1 reactions are more sensitive to this factor. So, an excellent leaving group suggests an S_N1 reaction.

It is possible to convert a bad leaving group into an excellent leaving group. For example, if we protonate an OH group, we convert it into a good leaving group:

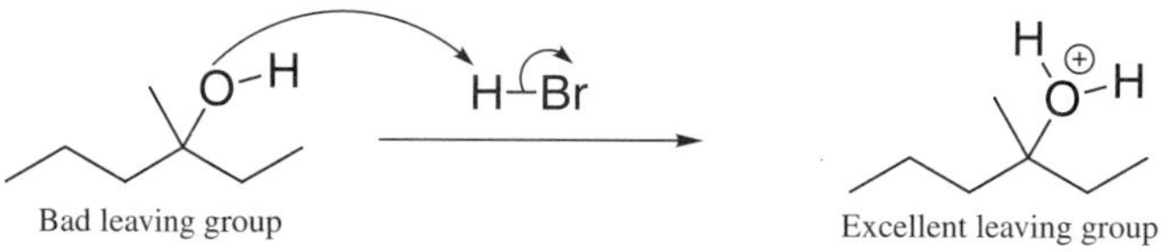

PROBLEMS For each compound below, determine whether the leaving group is excellent, good, or bad.

9.19   Answer: _____   9.20   Answer: _____

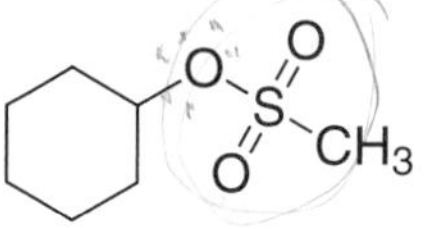

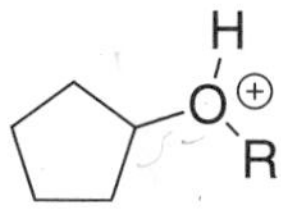

9.21 Answer: good   **9.22** Answer: bad

9.23 Answer: good   **9.24** Answer: bad

9.25 Using your answers to the last six problems, which of the six compounds above would you expect to be most likely to undergo an S_N1 reaction?

9.26 If you wanted to get the compound in Problem 9.24 to undergo an S_N1 reaction with chloride as the nucleophile, what would you need to do to the leaving group. What reagent would you use (to change the leaving group and to provide Cl^- at the same time)?

9.5   FACTOR 4—THE SOLVENT

So far, we have looked at the substrate, the nucleophile, and the leaving group. This takes care of all of the parts of the compounds that are reacting with each other. Let's summarize substitution reactions in a way that allows us to see this:

So, by talking about the substrate, the nucleophile, and the leaving group, we have covered almost everything. But there is one more thing to take into account. What solvent are these compounds dissolved in? It can make a difference. Let's see how.

There is a really strong solvent effect that *greatly* affects the competition between S_N1 and S_N2, and here it is: polar aprotic solvents favor S_N2 reactions. So, what are polar aprotic solvents, and why do they favor S_N2 reactions?

Let's break it down into two parts: *polar* and *aprotic*. Hopefully, you remember from general chemistry what the term "polar" means, and you should also remember that "like dissolves like" (so polar solvents dissolve polar compounds, and nonpolar solvents dissolve nonpolar compounds). Therefore, we really need a polar solvent to run substitution reactions. S_N1 desperately needs the polar solvent to stabilize the carbocation, and S_N2 needs a polar solvent to dissolve the nucleophile. S_N1 certainly needs the polar solvent more than S_N2 does, but you will rarely see a substitution reaction in a nonpolar solvent. So, let's focus on the term aprotic.

Let's begin by defining a protic solvent. We will need to jog our memories about acid–base chemistry. Recall that in Chapter 3 we talked about the acidities of protons (these are hydrogen atoms without the electrons, symbolized by H^+), and we saw that protons can be pulled off of a compound if the compound can stabilize the negative charge that develops when H$^+$ goes away. A protic solvent is a solvent that has a proton connected to an electronegative atom (for example, H$_2$O or EtOH). It is called protic because the solvent can serve as a source of protons. In other words, the solvent can give a proton because the solvent can stabilize a negative charge (at least a little bit). So what is an aprotic solvent?

Aprotic means that the solvent does *not* have a proton on an electronegative atom. The solvent can still have hydrogen atoms, but none of them are connected to electronegative atoms. The most common examples of polar aprotic solvents are acetone, DMSO, DME, and DMF:

Acetone

Dimethylsulfoxide (DMSO)

Dimethoxyethane (DME)

Dimethylformamide (DMF)

There are, of course, other polar aprotic solvents. You should look through your textbook and your class notes to determine if there are any other polar aprotic solvents that you will be expected to know. If there are any more, you can add them to the drawing above. You should learn to recognize these solvents when you see them.

So why do these solvents speed up the rate of S$_N$2 reactions? To answer this question, we need to talk about a solvent effect that is usually present when we dissolve a nucleophile in a solvent. A nucleophile with a negative charge, when dissolved in a polar solvent, will get surrounded by solvent molecules in what is called a *solvent shell:*

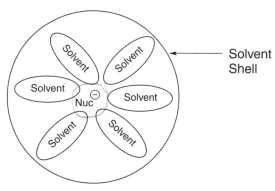

This solvent shell is in the way, holding back the nucleophile from doing what is supposed to do (go attack something). For the nucleophile to do its job, the nucleophile must first shed this solvent shell. This is always the case when you dissolve a nucleophile in a polar solvent, *except* when you use a polar aprotic solvent.

Polar aprotic solvents are not very good at forming solvent shells around negative charges. So if you dissolve a nucleophile in a polar aprotic solvent, the nucleophile is said to be a "naked" nucleophile, because it does not have a solvent shell. Therefore, it does not need to first shed a solvent shell before it can react with something. It never had a solvent shell to begin with. This effect is drastic. As you can imagine, a nucleophile with a solvent shell is going to spend most of its existence with the solvent shell, and there will be only brief moments every now and then when it is free to react. By allowing the nucleophile to react all of the time, we are greatly speeding up the reaction. S_N2 reactions performed with nucleophiles in polar aprotic solvents occur about 1000 times faster than those in regular protic solvents.

Bottom line: Whenever a solvent is indicated, you should look to see if it is one of the polar aprotic solvents listed above. If it is, it is a safe bet that the reaction is going to be S_N2.

EXERCISE 9.27 Predict whether the reaction below will occur via an S_N2 or an S_N1 mechanism:

Answer We look at the substrate and we see that it is secondary. That doesn't tell us much, so we move on to the nucleophile and we see that it is a moderate nucleophile (not strong and not weak); that doesn't tell us much either. We then look at the LG and see that it is a good LG (not excellent and not bad), which that doesn't tell us anything either. Finally, we look at the solvent. This is a polar aprotic solvent, and that serves as the tie breaker. S_N2 wins.

PROBLEM 9.28 Go back to the list of polar aprotic solvents, study the list, and then try to copy the list here without looking back.

9.6 USING ALL FOUR FACTORS

Now that we have seen all four factors individually, we need to see how to put them all together. When analyzing a reaction, we need to look at all four factors and make a determination of which mechanism, S_N1 or S_N2, is predominating. It may not be just one mechanism in every problem. Sometimes both mechanisms occur and it is difficult to predict which one predominates. Nevertheless, it is a lot more common to see situations that are obviously leaning toward one mechanism over the other. For example, it is clear that a reaction will be S_N2 if we have a primary substrate with a strong nucleophile in a polar aprotic solvent. On the flipside, a reaction will clearly be S_N1 if we have a tertiary substrate with a weak nucleophile and an excellent leaving group.

Your job is to look at all of the factors and make an informed decision. Let's put everything we saw into one chart. Review the chart. If there are any parts that do not make sense, you should return to the section on that factor and review the concepts.

Substrate	Nucleophile	Leaving group	Solvent
1°—Only S_N2, No S_N1	Strong—S_N2	Bad—Neither	Polar aprotic—S_N2
2°—Both	Moderate—Both	Good—Both (but more S_N2)	
3°—Only S_N1 No S_N2	Weak—S_N1	Excellent—S_N1	

EXERCISE 9.28 For the reaction below, look at all of the reagents and conditions, and determine if the reaction will proceed via an S_N2 or an S_N1, or both or neither.

Answer The substrate is primary, which immediately tells us that it needs to be S_N2. On top of that, we see that we have a strong nucleophile, which also favors S_N2. The LG is good, which doesn't tell us much. The solvent is not indicated. So, taking everything into account, we predict that the reaction follows an S_N2 mechanism.

PROBLEMS For each reaction below, look at all of the reagents and conditions, and determine if the reaction will proceed via an S_N2 or an S_N1, or both or neither.

9.29

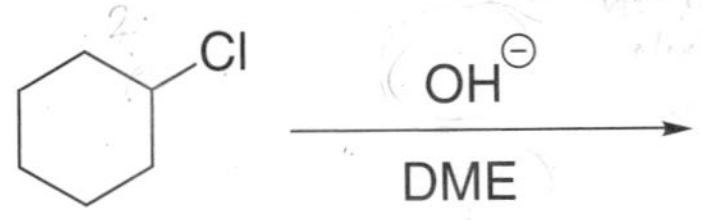

Cl / OH⁻ / DME → SN₂ ✓

9.30

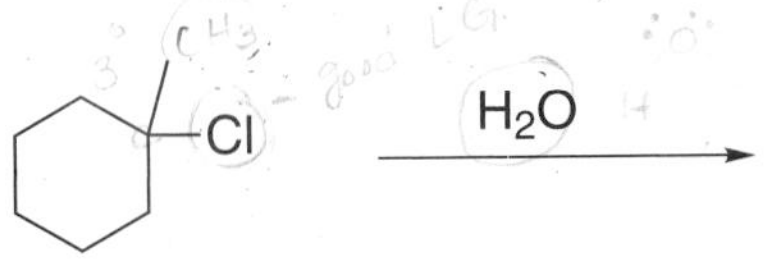

Cl / H₂O → SN₁

9.31

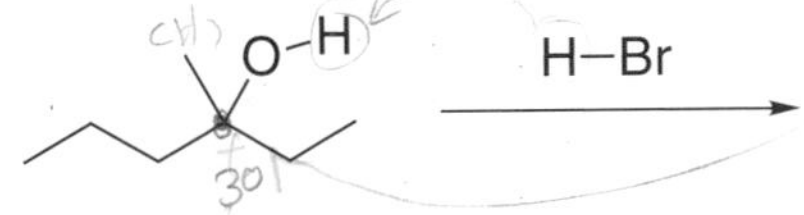

O–H / H–Br → SN₁

9.32

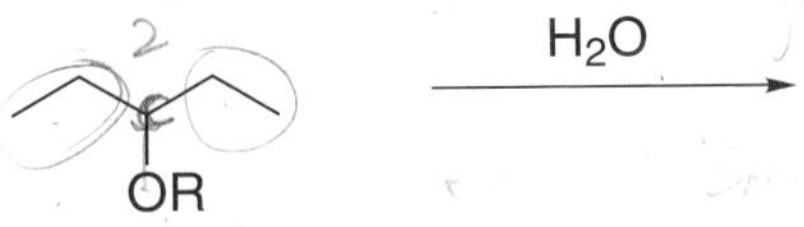

OR / H₂O → SN₁   either   neither

9.33

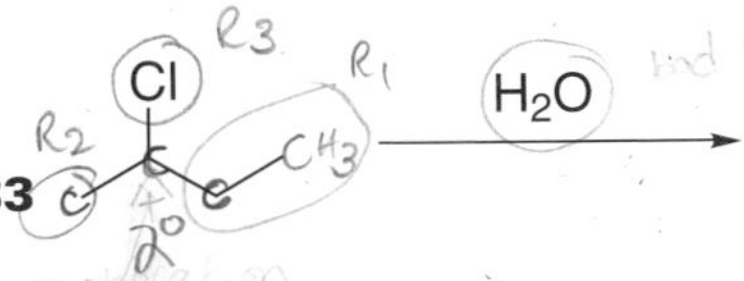

Cl / H₂O → SN₁, SN both?

9.34

O–S(=O)(=O)–O–CH₃ / ROH → SN₁ ✓

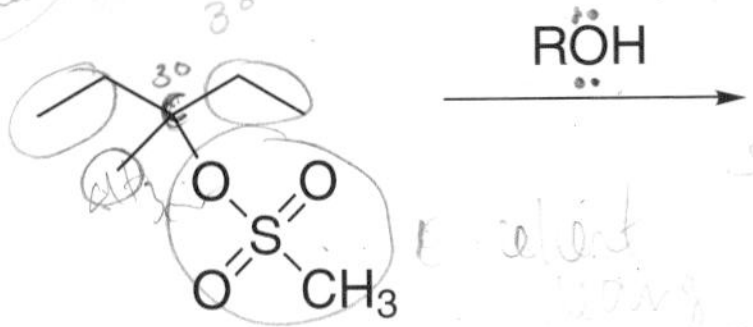

9.7 SUBSTITUTION REACTIONS TEACH US SOME IMPORTANT LESSONS

S_N1 and S_N2 reactions produce almost the same products. In both reactions, a leaving group is replaced by a nucleophile. In fact, the only difference in products between S_N1 and S_N2 reactions arises when you have a stereocenter where the leaving group is. In this situation, the S_N2 mechanism will invert the stereocenter, while the S_N1 mechanism will produce a racemic mixture. That's the only difference—the configuration of one stereocenter. And if there is no stereocenter, then there is no difference in outcome at all between S_N1 or S_N2. It seems like a lot of work to go through to determine the configuration of one stereocenter (which matters only some of the time).

So the obvious question is, why did we go through all of that trouble to learn how to determine whether a reaction is S_N1 or S_N2? There are many answers to this question, and it is important to spend some time on this, because it will help frame the rest of the course for you. Let's go through some answers one at a time.

First we learned the important concept that everything is located in the mechanisms. By understanding the mechanisms completely, everything else can be justified based on the mechanisms. All of the factors that influence the reaction can be understood by carefully examining the mechanism. This is true for every reaction you will see from now on. Now you have had some practice thinking this way.

Next we learned that there are multiple factors at play when analyzing a reaction. Sometimes the factors can all be pointing in the same direction, while at other times the factors can be in conflict. When they are in conflict, we need to weigh them and decide which factors win out in determining the path of the reaction. This concept of competing factors is a theme in organic chemistry. The experience of going through S_N1 and S_N2 mechanisms has prepared you for thinking this way for all reactions from now on.

Finally we learned that if we analyze the first factor (substrate), we will find two effects at play: electronics and sterics. We saw that S_N2 reactions require primary or secondary substrates because of sterics—it is too crowded for the nucleophile to attack a tertiary substrate. On the other hand, S_N1 reactions did not have a problem with sterics, but electronics was a bigger issue. Tertiary was the best, because the alkyl groups were needed to stabilize the carbocation.

These two effects (sterics and electronics) are major themes in organic chemistry. Much of what you learn in the rest of the course can be explained with either an electronic or a sterics argument. The sooner you learn to consider these two effects in every problem you encounter, the better off you will be. Electronics is usually the more complicated effect. In fact, the other three factors that we saw (nucleophile, leaving group, and solvent effects) were all electronic arguments. Once you get the hang of the kinds of electronic arguments that are generally made, you will begin to see common threads in all of the reactions that you will encounter in this course.

Don't get me wrong—it is very important to be able to predict whether a stereocenter gets inverted or not when a substitution reaction takes place. That alone would have been enough of a reason to learn all of the factors in this chapter. But I also want you to keep your eye on some of the "bigger picture" issues. They will help you as you move through the course.

CHAPTER *10*

ELIMINATION REACTIONS

In this chapter, we will explore elimination reactions in the same way that we explored substitution reactions. We begin with the mechanisms for E1 and E2 reactions, and then we move on to the factors that help us determine in each case which mechanism predominates. There is one big difference between the last chapter and this chapter. In the last chapter, most of the information was given to you, and there was very little to look up in other sources (your textbook, your class notes, etc.). But now you know how important mechanisms are, you know that mechanisms explain everything, you know how to analyze different factors that affect reactions, and so on. So in this chapter, YOU are going to provide the key information, by filling it in the appropriate places.

Don't worry. It will be a very interactive process. I will tell you what to look up and where to draw the information. We will go through what you need to do step by step. Let's see where we are in the grand scheme of things.

We are in the second part of a four-phase program to get you to the point where you can study without explicit instructions. Here are the four phases:

1. With substitution reactions, you were given all of the information so that you could see how to go through the process of analyzing each and every factor based on the mechanisms.

2. In this chapter, you will find all of the mechanisms and all of the factors yourself, but you will be told every step of the way what you should do.

3. In the next chapter (addition reactions), you will be asked to draw the mechanisms and record the important factors by yourself, without too much help.

4. Finally, you will record all of the information at the end of Chapter 8, which records every mechanism that you encounter. At this point, you will have the tools you need to study every reaction that you see in the rest of the course. You will know how to look at the mechanisms. You will know how to look for the factors that determine regiochemistry and stereochemistry. You will keep a record (in Chapter 8) of every reaction you learn, and you will know what you should focus on when you study.

So we are in the second part of this plan right now. Let's look at elimination reactions.

10.1 MECHANISMS (E1 AND E2)

We need to do the same analysis that we did for substitution reactions. So before we go any further, I recommend that you review Section 9.1 in the previous chapter. Do that now and then come back to here.

You need to start by drawing the mechanisms of the E1 and E2 reactions. Look in your class notes and in your textbook and then draw the mechanisms here:

PROBLEM 10.1 Draw the mechanism of the E2 reaction.

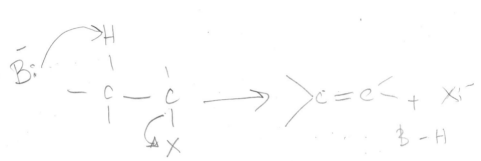

PROBLEM 10.2 Explain why the number "2" is used in the E2 mechanism. You will find that the reason is the same as the reason given for the "2" in S_N2.

PROBLEM 10.3 Draw the mechanism of the E1 reaction:

PROBLEM 10.4   Explain why the number "1" is used in the E1 mechanism. You will find that the reason is the same as the reason given for the "1" in S_N1.

_____

_____

Notice that the E1 mechanism goes through a carbocation intermediate. The formation of this intermediate will help explain how each of the factors plays a role. There are four factors to consider, just like in substitution reactions. And the four factors are directly analogous to the factors we have already seen. We will explore those factors in the upcoming sections.

10.2   FACTOR 1—THE SUBSTRATE

Just as in substitution reactions, the first factor is the substrate. Go through your class notes and your textbook to determine what to expect for primary, secondary, and tertiary substrates. For each type of substrate, you need to find out if E2 is favored, if E1 is favored, or if they are both favored. Use the mechanisms to help you understand why. Then use the charts below to record the trends for E1 and E2 (just like the charts we used to compare the trends for S_N2 and S_N1 in Section 9.2 of the previous chapter):

PROBLEM 10.5   Fill in the trends for E1 and E2 reactions, comparing relative rates for primary, secondary, and tertiary substrates.

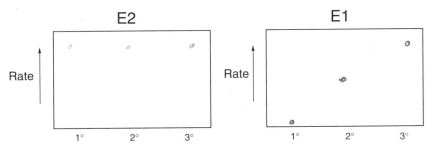

Keep in mind when filling in the E2 chart, that you don't have the same steric concerns in tertiary substrates that you had in S_N2 reactions. The base does not need to get in and attack the carbon. It only has to pull off a proton. Sterics is no longer an effect in this case, and you can have E2 reactions on tertiary substrates:

S_N2 _cannot_ happen   E2 _can_ happen

PROBLEMS For each compound below, determine (based on its structure—i.e., primary, secondary, or tertiary) whether you expect an E2 mechanism or whether both E1 and E2 can happen (remember that E2 can happen in all substrates):

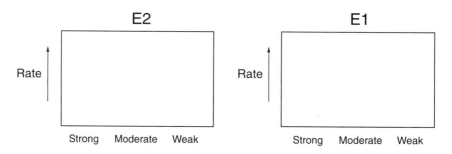

10.6 _____ E₂   both _____   **10.7** _____ E₂   E₁   both. _____

10.8 _____ E₂ _____   both   **10.9** _____ E₁ neither _____

10.3   FACTOR 2—THE BASE

You should begin by reviewing Section 9.3 of the last chapter. In that section, we saw that factor 2 for substitution reactions was the nucleophile. In elimination re-actions, we are looking at a *base* instead. Remember that the difference between bases and nucleophiles is function (we saw this in Chapter 8, Section 8.3—review this if you don't remember it). Most reagents can function as either a nucleophile or a base. For example, hydroxide can act as a nucleophile in an S_N2 reaction or it can act like a base in an E2 reaction. In fact, this is the critical difference between substitution and elimination reactions—does the reagent function as a nucleophile and attack or does it function as a base and pull off a proton? So we are looking at base strength instead of nucleophile strength. Go over your class notes and your textbook and try to fill in the charts below that show the trends for factor 2, base strength:

PROBLEM 10.10 Fill in the trends for E1 and E2 reactions, comparing relative rates based on base strength:

These charts should look the same as the charts for factor 2 of substitution reactions. If you cannot figure it out without looking, you can always turn back to the last chapter, to Section 9.3.

PROBLEMS Using the charts above, determine whether the following bases would favor an E2 or an E1.

10.11 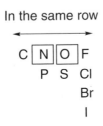 Answer: _____

10.12 Answer: _____

10.13 Answer: _____

10.14 H_2O Answer: _____

We saw in Chapter 8 (Section 8.3) that the difference between bases and nucleophiles is function. We even saw the difference between the terms nucleophilicity (which is a kinetic concept) and basicity (which is a thermodynamic concept). But there is another important difference between the strengths of bases and nucleophiles that you must know. Students rarely see this difference, and it causes them much unnecessary anguish when doing problems that involve substitution and elimination reactions at the same time (we will do problems like this in Chapter 12). Let's avoid the anguish by clearing up the difference now.

Students generally assume that a stronger base is a stronger nucleophile. This is not always true. Basicity and nucleophilicity do not always parallel each other. Let's begin by seeing when they *do* parallel each other.

When comparing atoms *in the same row* of the periodic table, basicity and nucleophilicity *do* parallel each other:

In the same row

C N O F
P S Cl
Br
I

For example, let's compare NH_2^- and OH^-. The difference between these two compounds is the atom bearing the charge (O vs. N). We saw in Chapter 3 (when we look at the factors determining charge stability) that oxygen, being more electronegative than nitrogen, can stabilize a charge better than nitrogen can. Therefore, OH^- will be more stable than NH_2^-, so NH_2^- will be a stronger base. As it turns out, NH_2^- will also be a stronger nucleophile than OH^-, because basicity and nucleophilicity parallel each other when comparing atoms in the same row of the periodic table.

When comparing atoms *in the same column* of the periodic table, basicity and nucleophilicity do not parallel each other:

In the same column

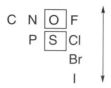

For example, let's compare OH$^-$ and SH$^-$. The difference between these two compounds is the atom bearing the charge (O vs. S). We already saw in Chapter 3 that sulfur, being larger than oxygen, can stabilize a charge better than oxygen can (remember that size is more important than electronegativity when comparing atoms in the same column). Therefore, SH$^-$ will be more stable than OH$^-$, so OH$^-$ will be a stronger base. However, SH$^-$ is a better nucleophile than OH$^-$, even though OH$^-$ is a better base than SH$^-$. Why?

Recall that basicity and nucleophilicity are different concepts. Basicity measures stability of the charge (a thermodynamic argument), whereas nucleophilicity measures how fast a nucleophile attacks something (a kinetic argument). When you have a large atom, like sulfur, an interesting effect comes into play. As the sulfur atom approaches an electrophile (a compound with δ^+), the electron density within the sulfur atom gets polarized, meaning that the electron density can move around.

This effect causes the sulfur to be drawn to the electrophile very quickly, so the rate of attack is very fast. Since *nucleophilicity* is a measure of how fast the nucleophile attacks, this effect makes the sulfur atom very nucleophilic. When you have a negative charge on oxygen, the electron density is not polarizable as in the sulfur atom, because the atom is too small. Hence, this effect is not present in the oxygen atom.

So if we want to measure basicity, we are looking at stability. Sulfur can better stabilize the charge (because it is larger), which means that it is more stable. Therefore, OH$^-$ is a stronger base than SH$^-$. However, when we measure nucleophilicity, we see that the sulfur atom is polarizable, and so it is an excellent nucleophile. We see that oxygen is a better base than sulfur, but sulfur is a better nucleophile than oxygen. In fact, this effect (polarizability for large atoms like sulphur) is such a strong effect, that compounds containing sulfur will act almost exclusively as nucleophiles, and not as bases. This is where we get to the bottom line.

Bottom line: sulfur nucleophiles will function as nucleophiles and not as bases. For the same reason, halides (Cl$^-$, Br$^-$, I$^-$, which are all very large and very polarizable) will also function as nucleophiles and not as bases. So when you see one of these nucleophiles, you do not need to worry about elimination reactions—you will get *only* substitution reactions. It is very common to see the halides being used as nucleophiles, so it is very helpful to know that you do not need to worry about elimination reactions when you see a halide as the reagent. In Chapter 12, we will see how important this will be to determine whether substitution reactions or elimination reactions predominate.

10.4 FACTOR 3—THE LEAVING GROUP

This is exactly the same factor as we saw in substitution reactions. The trends are identical. In fact, if you compare the first step of an S_N1 mechanism with that of an E1 mechanism, they are identical. Go back and review Section 9.4 of the previous chapter. Then come back to here and fill in the charts below.

PROBLEM 10.15 Fill in the trends for E1 and E2 reactions, comparing relative rates based on stability of the leaving group:

E2

Rate

Excellent   Good   Bad

E1

Rate

Excellent   Good   Bad

PROBLEMS For each compound below, determine whether the leaving group is excellent, good, or bad. Based on that, and using the charts above, determine if the leaving group favors E2, E1, both, or neither.

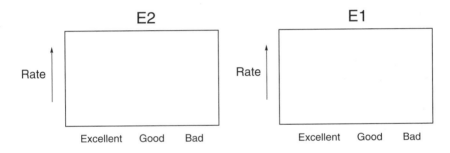

good   both

10.16   Answer:

E₁

10.17   Answer:____

10.18   Answer:____   both

10.19   Answer:____   neither

10.20   Answer:____   E₁   good both

10.21   Answer:____   neither

10.5   FACTOR 4—SOLVENT EFFECTS

Finally, explore any solvent effects that are listed in your textbook, if any are discussed (usually they are not discussed). Record what you find in the space below:

10.6   USING ALL OF THE FACTORS

Now you have gone through all of the factors yourself. Now you should summarize all of this information in the following chart (just like we did for substitution reactions). For each factor, indicate which mechanism is favored. The first column has been filled in for you:

Substrate	Base	Leaving group	Solvent
1°—E2 only	Strong—	Bad—	
2°—E2	Moderate—	Good—	
3°—E2 and E1	Weak—	Excellent—	

If you study the chart above (after you have filled it in), you will find that an E1 mechanism is very difficult to have. All of the conditions have to be just right—we need a tertiary substrate (so the leaving group can leave to form a stable carbocation), a weak base (so that E2 is not competing), and an excellent leaving group (so the leaving group can leave on its own). If any of these conditions are not met, than we more likely get an E2 reaction. So we will more commonly see E2 reactions, especially since E2 can occur very rapidly for tertiary substrates.

EXERCISE 10.22   Using the information that you constructed in the chart above, predict whether the following reaction will proceed via an E2 mechanism or an E1 mechanism:

Answer   The substrate is tertiary, so it could be E1 or E2. The base is a strong base (a negative charge that is not resonance stabilized), so the E2 mechanism will be faster.

PROBLEMS   For each of the reactions below, predict whether the reaction will proceed via an E2 mechanism, an E1 mechanism, or neither. For now, do not worry about drawing the products. We need to cover the next section before we can do that. Right now, just focus on determining which mechanism, if any, is possible.

10.23

10.24

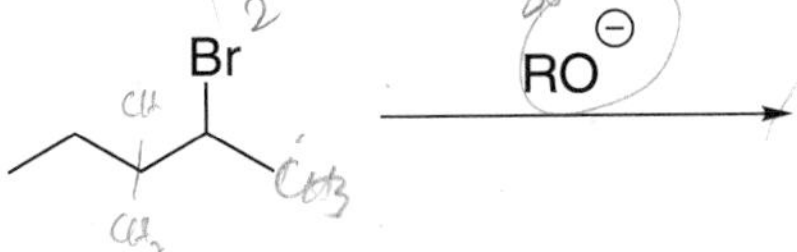

10.25

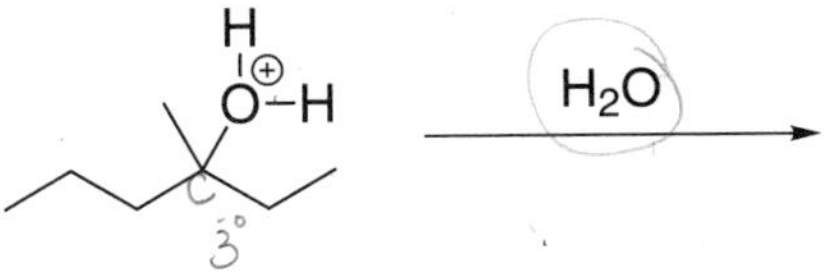

10.26

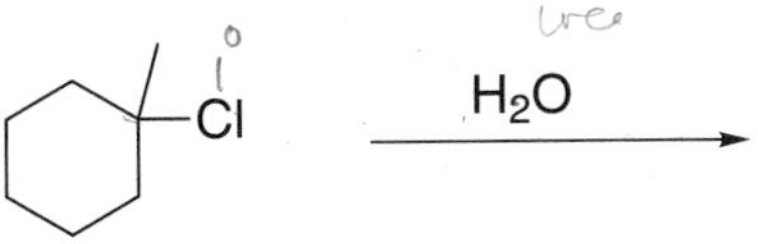

10.27

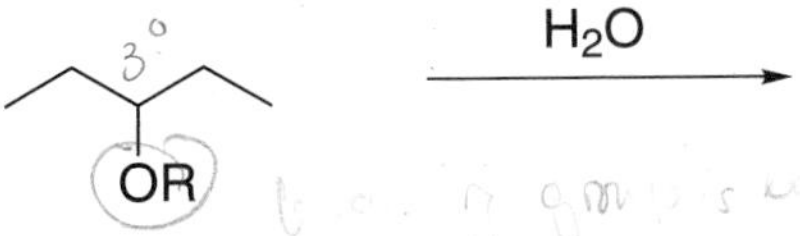

10.7 ELIMINATION REACTIONS— REGIOCHEMISTRY AND STEREOCHEMISTRY

We have mentioned many times that you need to think about the regiochemistry and stereochemistry of every reaction. We will now consider those issues for elimination reactions, beginning with regiochemistry.

Regiochemistry refers to where the reaction takes place. In other words, in what *region* of the molecule is the reaction taking place? When you eliminate H and X (where X is some leaving group), it is possible to form the double bond in different locations. Consider the following simple example:

Where does the double bond form? This is a question of regiochemistry. The way we distinguish between these two possibilities is by considering how many groups are attached to each double bond. Double bonds can have anywhere from 1 to 4 groups attached to them:

| Monosubstituted | Disubstituted | Trisubstituted | Tetrasubstituted |

So, if we look back at the reaction above, we find that the two possible products are monosubstituted and disubstituted double bonds. Whenever you have an elimination reaction where more than one possible double bond can be formed, we have names for the different products based on which one is more substituted and which one is less substituted. The more substituted product is called the Zaitsev product, and the less substituted product is called the Hoffmann product. Usually you get the Zaitsev product, but under special circumstances you can get the Hoffman product. If you use a strong, sterically hindered base, you can form the Hoffman product.

PROBLEM 10.28 Search through your textbook, find the section that covers formation of Hoffmann products, and then draw the structures of the sterically hindered bases that your textbook shows you.

Whenever you see these bases, you should immediately recognize that they will give the Hoffman product.

PROBLEMS For each of the compounds below, draw the Zaitsev and Hoffman products of an E2 reaction.

10.29

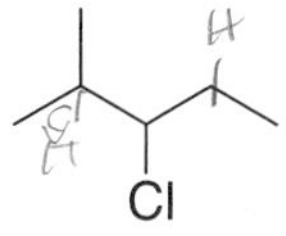

Zaitsev	Hoffmann

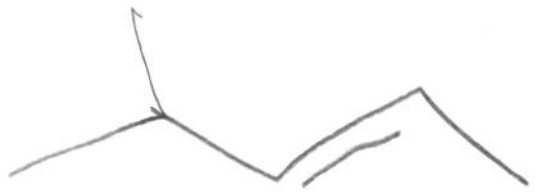

10.30

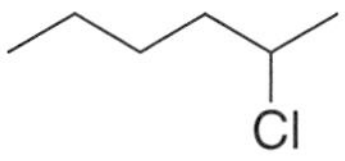

Zaitsev	Hoffmann

10.31

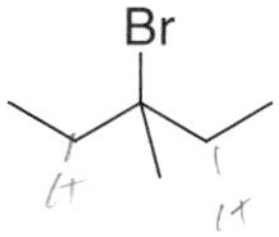

Zaitsev	Hoffmann

Now let's turn our attention to the stereochemistry of elimination reactions. E1 reactions go through an intermediate carbocation, so you lose stereospecificity. This means that if there are two possible stereoisomeric double bonds, you will get both of them:

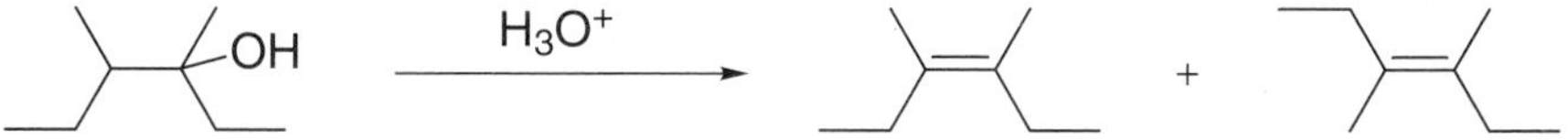

· In the reaction above, we know that the regiochemistry dictates that we form a tetrasubstituted double bond (rather than the other possibilities in this case). But now we need to decide whether we get both possible stereoisomers (cis and trans). E1 mechanisms will form both products.

But E2 reactions are stereospecific—the H and the LG must be antiperiplanar during the reaction, which defines which stereoisomeric double bond you get. This is best illustrated using Newman projections. If you draw a Newman projection, you can then easily rotate into the conformation that has the H and the LG antiperiplanar. This conformation then shows you which stereoisomer you get. Consider the following example.

EXERCISE 10.32   Predict the elimination product(s) of the following reaction:

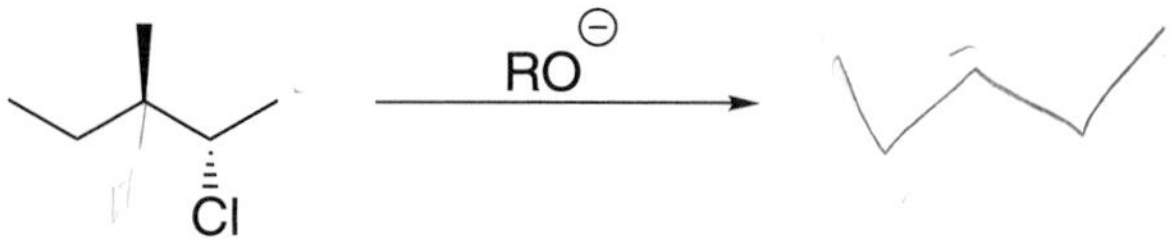

Answer   The regiochemistry of the reaction will be the formation of the more sub-stituted double bond, so the double bond forms to the left of the LG, rather than to the right.

We have a strong base, so we know the reaction will go through an E2 mech-anism, rather than an E1 mechanism. Therefore, the reaction will be stereospecific. So we draw the Newman projection:

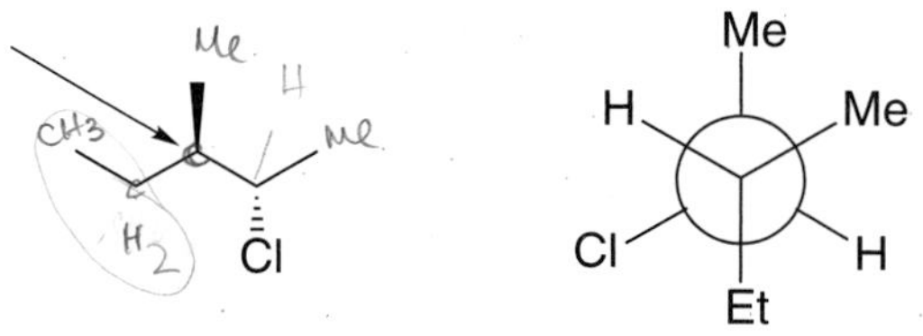

Next we need to draw rotate to the conformation that puts the Cl and H (on the front) carbons anti to each other:

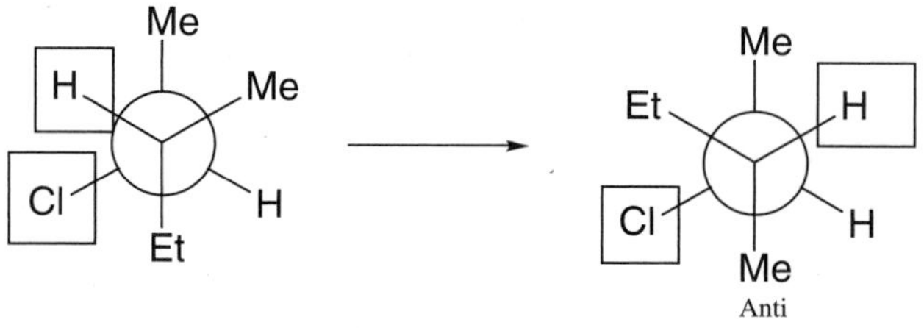

This is the conformation from which the reaction can take place. The double bond is being formed between the front carbon and the back carbon. This Newman projec-tion shows us whether we get cis or trans:

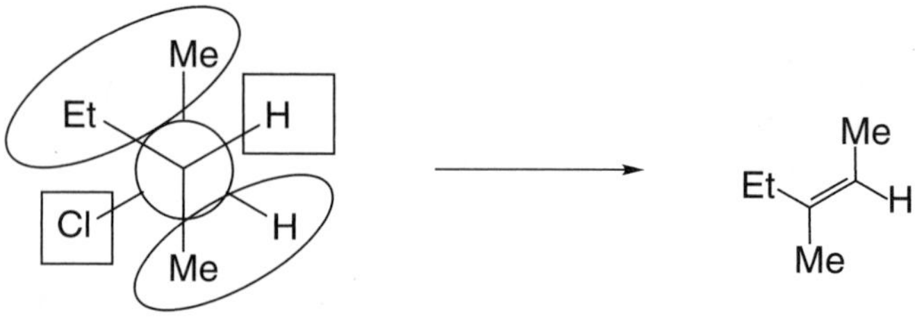

So the product is a trans double bond (in this case). We will not form the cis prod-uct because the E2 reaction is stereospecific.

You need to get into the habit of drawing Newman projections so that you can determine the stereoisomer that you get in an E2 reaction. If you are rusty on Newman projections, you should go back and review Sections 6.1 and 6.2 of Chapter 6. Then come back to here and try to use Newman projections to determine the stereochemistry of the following reactions.

PROBLEMS For each of the following compounds, predict what the product would be of an E2 reaction (assume the Zaitsev product and focus on stereochemistry):

10.33 _____ _____ 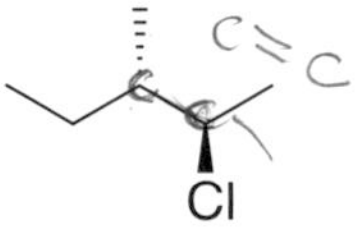 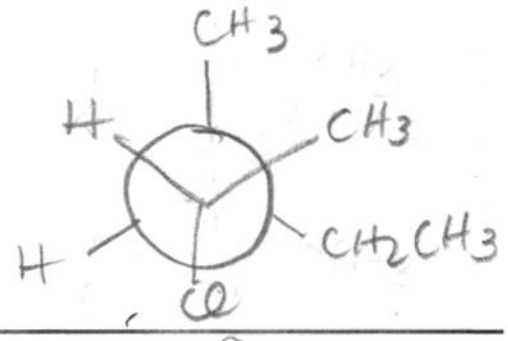
Newman projection   Final answer

10.34 _____ _____ 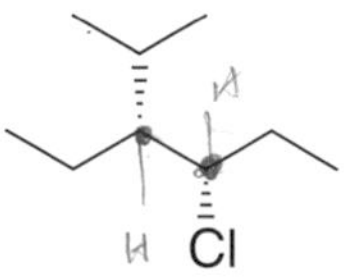 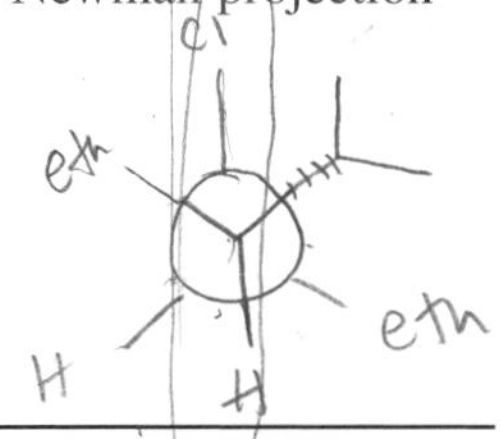 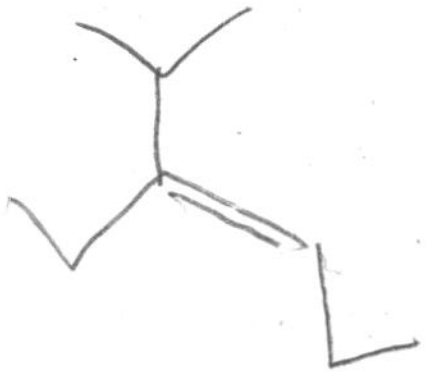
Newman projection   Final answer

10.35 _____ _____ 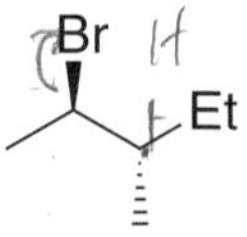 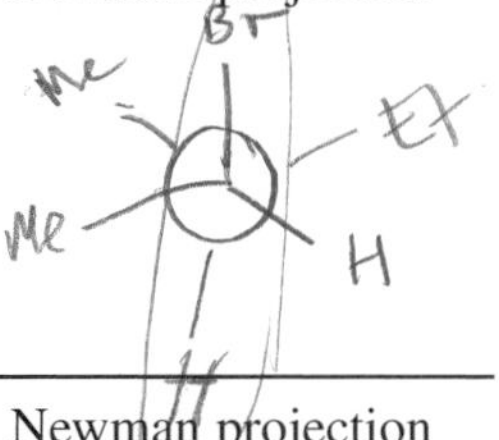 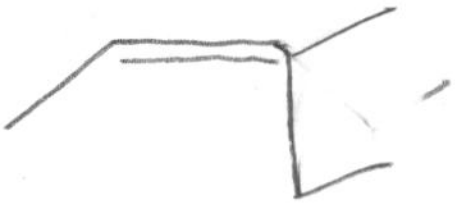
Newman projection   Final answer

10.36 _____ _____ 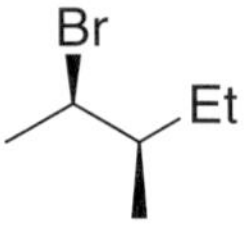 
Newman projection   Final answer

There are times when you can have substitution and elimination reactions occurring at the same time. In fact, this is often the case. In those situations, you need to determine which of the four mechanisms (S_N2, S_N1, E2, or E1) is predominant. We will see this in Chapter 12 (Section 12.3).

CHAPTER 11

ADDITION REACTIONS

We saw in the last chapter that we are in the middle of a four-phase program to get you to the point where you can study without explicit instructions. Let's review where we are in this process. Here are the four phases:

1. With substitution reactions, you were given all of the information so that you could see how to go through the process of analyzing each and every factor based on the mechanisms.

2. With elimination reactions, you were told how to find the information you need to understand elimination reactions (mechanisms, factors, etc.), and you recorded the information as you went along (in the form of charts and other drawings).

3. In this chapter, you will be asked to draw the mechanisms and record the important factors by yourself, without too much help.

4. Finally, when you have finished this chapter, you will record all of the information at the end of Chapter 8, which records every mechanism that you encounter. At that point, you will have the tools you will need to study every reaction that you see in the rest of the course. You will know how to look at the mechanisms. You will know how to look for the factors that determine regiochemistry and stereochemistry. You will keep a record (in Chapter 8) of every reaction you learn, and you will know what you should focus on when you study.

Now, let's get started.

We saw in the last two chapters that mechanisms are the keys to understanding everything. So it makes sense that we need to begin with mechanisms. For each reaction that you learn, you will need to focus on the mechanism. Once you understand that, you will be in great shape to understand the regiochemistry and stereochemistry.

In addition reactions, regiochemistry refers to where the two groups end up on the double bond. When we add two groups across a double bond (let's call them X and Y), then there are two possibilities:

Let's see a specific example.

Consider the addition reaction of HCl across a double bond:

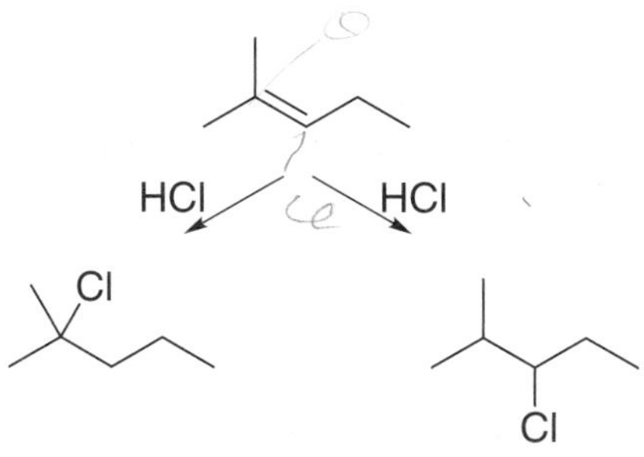

There are two possible ways to add the H and the Cl. Which product do we get?

One possibility would be to put the Cl on the more substituted carbon, and the other possibility would be to put the carbon on the less substituted carbon. If we put the Cl on the more substituted carbon, we call this a Markovnikov addition. If we put the Cl on the less substituted carbon, we call this an anti-Markovnikov addition. In this reaction, we get Markovnikov addition. But why? The answer is, of course, in the mechanism.

The first step in the mechanism involves the electrons of the double bond attacking the proton of HCl. There are two arrows necessary to make this happen. One arrow goes from the double bond to the proton, and the other arrow goes from the H-Cl bond to the Cl. There are two possible ways to do this:

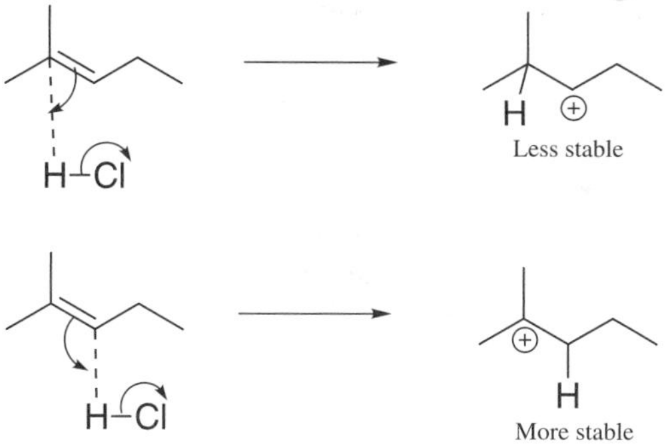

The difference between these two possibilities is where the H^+ goes in the first step. The product of each possibility is a carbocation. Recall that alkyl groups are electron donating, so the tertiary carbocation on the bottom will be more stable than the secondary carbocation on the top. Therefore, in the next step, the Cl^- will go wherever the carbocation is, which will be at the more substituted carbon:

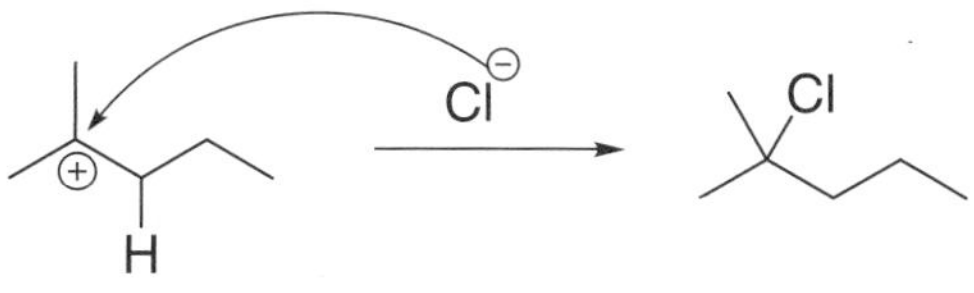

So we see that the final position of the chlorine is determined by carbocation stability, which becomes evident as we work through the mechanism. The mechanism for any reaction will help to explain the regiochemistry of that reaction.

For every addition reaction that you learn, you must always think about the regiochemistry of the reaction. Sometimes you will see that it is not an issue. For example, if you are adding H and H across a double bond, then regiochemistry is not an issue. It does not matter which carbon gets the first H and which carbon gets the second H. Either way, they both end up with an H. Similarly, if we add two OH groups across a double bond, regiochemistry also does not matter. Any time you add two of the same group across a double bond, you will not have to worry about the regiochemistry.

Now let's shift our attention to stereochemistry. In addition reactions, stereochemistry tells us how the groups added in 3D space. There are two possibilities: syn addition or anti addition. Let's look at some examples to better understand this.

Consider the addition of OH and OH across a double bond. We already saw that we don't need to worry about the regiochemistry of this reaction, because we are adding two of the same group. But what about the stereochemistry?

We are creating two new stereocenters, so there are four possibilities: RS, SR, RR, or SS. How many of them do we get? There are two sets of enantiomers: RR and SS represent one set of enantiomers, and RS and SR represent the other set of enantiomers. Do we get both sets as our products (meaning all four products), or do we only get one set of enantiomers? This depends on how the reaction took place.

If an addition reaction can take place only through a mechanism that allows a syn addition, then the two groups that we added must be on the same side of the double bond in the product. So we will get only that one set of enantiomers:

If a reaction can only go through an anti addition, then the two groups we added must be on opposite sides of the double bond. So we will get only the set of enantiomers where the groups are on opposite sides:

Sometimes, the reaction is not stereoselective. In other words, we get both syn and anti addition. So, we get all four products (both sets of enantiomers).

Each reaction will be different. Some will give only syn addition, some will give only anti addition, and others will not be stereoselective. For every addition reaction, you need to know the stereochemistry of the addition, and that information is contained within the mechanism.

In the example above, the addition of OH and OH across a double bond (using OsO_4 as a reagent) will always be a syn addition:

Once again, all of the stereochemical information for any reaction is contained within the mechanism. So your first step is to master the mechanisms.

Bottom line: How should you study addition reactions? For every addition reaction that you encounter, you must draw the mechanism first. Once you completely understand it, then you can look for the stereochemistry and regiochemistry and try to justify them based on the mechanism. Then you will be in a position to understand any of the factors that your textbook mentions about that reaction. Those factors will often help you determine when and how quickly the reaction occurs. There will usually be fewer factors than we saw in substitution and elimination reactions. Usually only one or two factors will be covered on any reaction (if even that). You should then turn to the end of Chapter 8 and summarize this information for each reaction. You will record the mechanism and the key information regarding stereochemistry and regiochemistry. If you repeat this process for every reaction that you learn (not just addition reactions, but all reactions), then you will be in really good shape.

A sample entry is shown below:

Reaction type	Stereochemistry	Regiochemistry
Addition of Br_2 across a double bond	Anti addition	NA—Adding two of the same group across the double bond

Don't wait until the day before the exam to do this for every reaction. It will take too long. You must keep up with it every day, every time you learn a new reaction.

CHAPTER 12

PREDICTING PRODUCTS

12.1 GENERAL TIPS FOR PREDICTING PRODUCTS

In the last several chapters, we saw the power of mechanisms in explaining the regiochemistry and stereochemistry of every reaction. A solid understanding of these concepts gives us the power to predict the products of a reaction. That is the subject of this chapter.

Generally, there are not many steps involved. You need to ask three questions whenever you are trying to determine the product of a reaction:

1. What kind of reaction is taking place?
2. What is the regiochemistry of that reaction?
3. What is the stereochemistry of that reaction?

By studying the mechanisms of every reaction and recording them at the end of Chapter 8, you will become familiar with these three aspects of every reaction and you will find it simple to predict products. Let's see an example of how this works.

EXERCISE 12.1 Predict the product of the following reaction:

$$(1)\ BH_3 / THF$$
$$(2)\ H_2O_2 / OH^{\ominus}$$

Answer If you have already learned this reaction and if you have recorded the mechanism at the end of Chapter 8, then you will already know the regiochemistry and stereochemistry of this reaction. Let's answer all three questions:

1. What kind of reaction is taking place?
   —Addition of H and OH across a double bond
2. What is the regiochemistry of this reaction?
   —The OH goes on the less substituted carbon (anti-Markovnikov addition)
3. What is the stereochemistry of this reaction?
   —syn addition

Now that we have answered all three questions, we are ready to draw the products (don't forget to draw both enantiomers that you get from syn addition):

If you fail to ask all three questions, then you will not get the product correct. Many students fail to ask questions 2 or 3 (or sometimes both). Clearly, the answer is correct only if you get the stereochemistry and the regiochemistry correct.

12.2   GETTING PRACTICE

This section offers one approach for practicing predicting products: For every reaction that you learn, we have already said that you should record it at the end of Chapter 8. Now we need to start a new list, right here in this chapter. Every time you learn a new reaction, you should write the reaction down using the arrows on the pages that follow, but do not show the mechanism or the product. Just show the starting compound and the reagents, like this:

As you learn more and more reactions, this list will grow. With every five new reactions, you should photocopy all of the reactions that you have recorded here. Then, start filling in the products on the photocopy. If you cannot fill them all in, go back to the end of Chapter 8 where you first recorded the reaction, and see how to answer the problem. If you did not fill out the end of Chapter 8, then look through your textbook and your class notes to determine the answer to all three questions (what kind of reaction? regiochemistry? and stereochemistry?). Repeat this procedure whenever you have entered five new reactions.

If you keep up with this exercise as the course progresses, you will be in very good shape for predicting products. The hardest challenge that you will face is keeping up with the work and not waiting until the night before the exam. If you wait (as most students do), you will find it very difficult to spend the time that it takes to master this material. Don't make that mistake. The secret to success in this course is to do a little bit every night (rather than cramming on the night before the exam).

Begin your list on the next page.

Remember not to fill in the products or the mechanisms. For each reaction, just draw the starting material in front of the arrow and the reagents above the arrow. Leave the space for the product empty. You will fill in the products when you photocopy these pages:

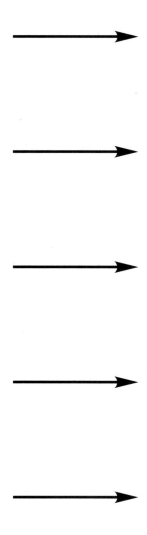

Now photocopy this page, and try to fill in the products on your photocopied page.

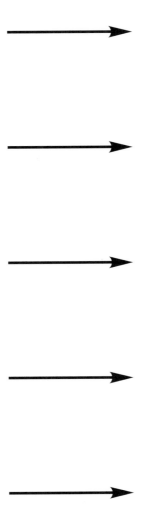

Now, photocopy this page again, and fill in the products for every reaction on this page.

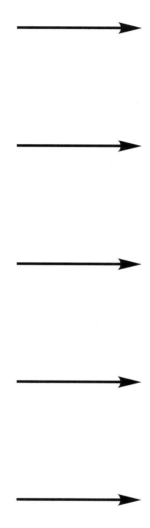

Now photocopy this page AND the previous page, and fill in all of the products.

Now photocopy this page AND the previous page, and fill in all of the products.

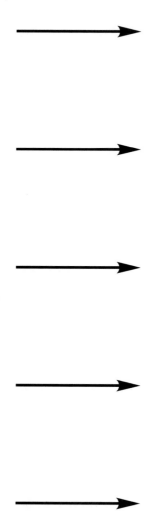

Now photocopy this page AND the previous pages, and fill in all of the products.

Now photocopy this page AND the previous pages, and fill in all of the products.

Now photocopy this page AND the previous pages, and fill in all of the products.

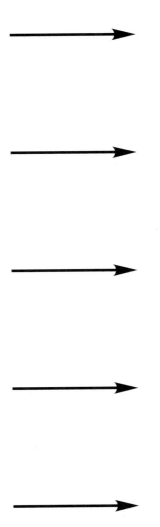

Now photocopy this page AND the previous pages, and fill in all of the products.

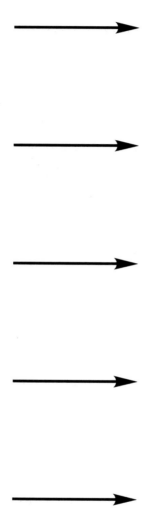

Now photocopy this page AND the previous pages, and fill in all of the products.

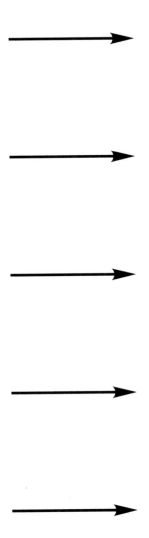

Now photocopy this page AND the previous pages, and fill in all of the products.

If you cover more than 30 reactions and need more space to continue, then you can just use a regular piece of paper to keep your list going.

12.3 SUBSTITUTION VERSUS ELIMINATION REACTIONS

Predicting products can be challenging when you have to consider substitution and elimination reactions simultaneously. So far, we have seen substitution and elimination reactions separately. But now the truth comes out—substitution and elimination reactions are generally in competition with each other. To predict the products properly, you need to compare all factors for substitution *and* elimination reactions, and you then need to decide which of the four mechanisms predominates (S_N1 or S_N2 or E1 or E2).

The method for predicting products is still the same. We ask the same three questions:

1. What kind of reaction is taking place?
2. What is the regiochemistry of that reaction?
3. What is the stereochemistry of that reaction?

The reason these problems get a bit more challenging is that question 1 (what kind of reaction) is a bit more tedious to work through. You need to analyze all four factors: the substrate, the reagent (which could be a nucleophile or a base), the leaving group, and the solvent. Once you have analyzed all four factors, you will be able to determine which mechanisms are operating. Then for each mechanism that is operating, you need ask the remaining two questions (regiochemistry and stereochemistry) to draw the products.

Let's quickly review the regiochemistry and stereochemistry of each of the four mechanisms:

S_N1

    Stereochemistry = racemization

    Regiochemistry = not applicable (the nucleophile simply attacks at the carbon connected to the leaving group)

    For example,

S_N2

    Stereochemistry = inversion of configuration

    Regiochemistry = not applicable (the nucleophile simply attacks at the carbon connected to the leaving group)

For example,

E1

Stereochemistry = if there are cis/trans isomers, you will get both
Regiochemistry = form the more substituted double bond (Zaitsev product)

For example,

E2

Stereochemistry = if there are cis/trans isomers, you will get the one determined by the H and LG being antiperiplanar (draw Newman projections)
Regiochemistry = form the more substituted double bond (Zaitsev product). If you are using a strong, sterically hindered base, then form the
less substituted double bond (Hoffmann product)

For example,

It is OK to have more than one product for a reaction. For example, it is OK
to determine that S$_N$2 and E2 are competing and that you get products from both
mechanisms at the same time. There is nothing wrong with that. There does not have
to be only one answer. But there are a few extra pieces of information that will help
you determine if there is one or more than one mechanism operating:

1. Some reagents are excellent nucleophiles, but are not good bases. Some
   reagents are excellent bases, but are not good nucleophiles. We discussed
   this in Chapter 10 (Section 10.3). If you do not understand how a reagent
   can be a good nucleophile and not a good base, go back to this section and
   review it. You should be familiar with certain reagents:

   - Nucleophiles (that do *not* act as bases): F$^-$, Cl$^-$, Br$^-$, I$^-$, CN$^-$, RS$^-$,
     RSH. When you see these reagents, you do not need to worry about elim-
     ination reactions.
   - Bases (that do not function as nucleophiles): H$^-$. When you see this
     reagent, you do not need to worry about substitution reactions.
   - Most other reagents that you will see can function as either a base or a
     nucleophile.

2. Look at the temperature, if it is given. High temperature (like 100°C or higher) will favor elimination products. Low temperature (like room temperature) will favor substitution products.

3. Bimolecular reactions are generally faster than unimolecular reactions. So, all things being equal, S_N2 will be faster than S_N1, and E2 will be faster than E1.

Let's do some examples and see how it works.

EXERCISE 12.2   Predict the products of the following reaction:

Answer   We need to ask our three questions:

1. What kind of reaction is taking place?

We see that we have a substrate with a leaving group, so we need to consider substitution and elimination reactions. The reagent is a good nucleophile, but it is not a good base, so we only need to worry about substitution reactions. We need to determine whether we have S_N1 or S_N2. Let's go through all four factors:

- The substrate is secondary, so that doesn't help us much.
- The nucleophile is pretty good (it has negative charge) so that favors S_N2.
- The leaving group is good (but not excellent and not bad) so that doesn't tell us much.
- The solvent is a polar aprotic solvent. This tells us that we have an S_N2 reaction.

Now we know what reaction is taking place (S_N2), so we ask our last two questions:

2. What is the regiochemistry?

Not applicable. The nucleophile attacks at the carbon that bears the leaving group.

3. What is the stereochemistry?

Inversion.

Now we are ready to draw our product. We replace the Br with a CN group and we invert the stereocenter:

This reaction had only one product. There are times when you can get more than one product.

PROBLEMS Predict the products of each of the reactions below.

12.3

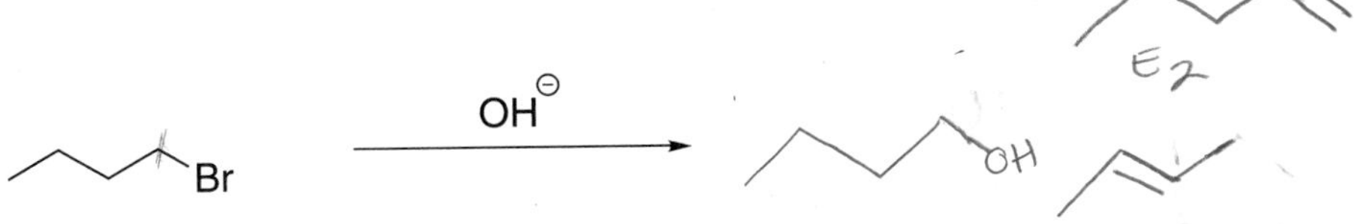

(structure: CH₃CH₂CH₂CH₂-Br) $\xrightarrow{\text{OH}^{\ominus}}$

E2

SN2

12.4

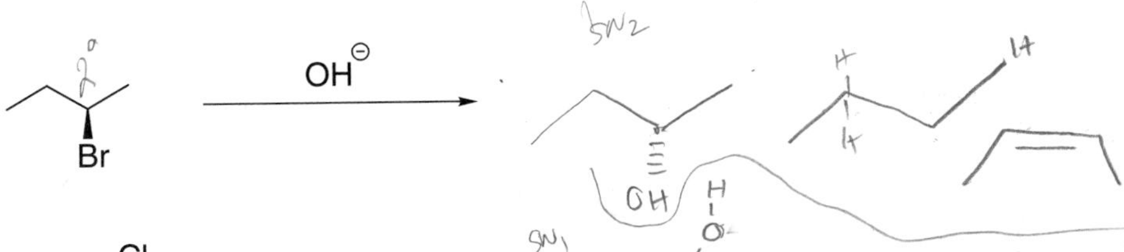

(structure with Br) $\xrightarrow{\text{OH}^{\ominus}}$

SN1

OH H

12.5

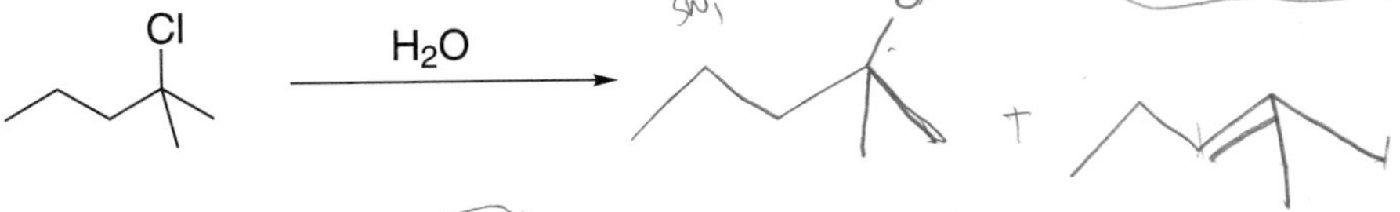

(structure with Cl) $\xrightarrow{\text{H}_2\text{O}}$

12.6

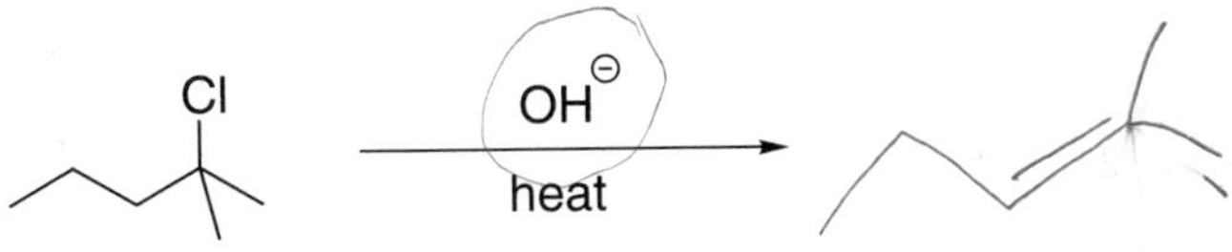

(structure with Cl) $\xrightarrow[\text{heat}]{\text{OH}^{\ominus}}$

12.7

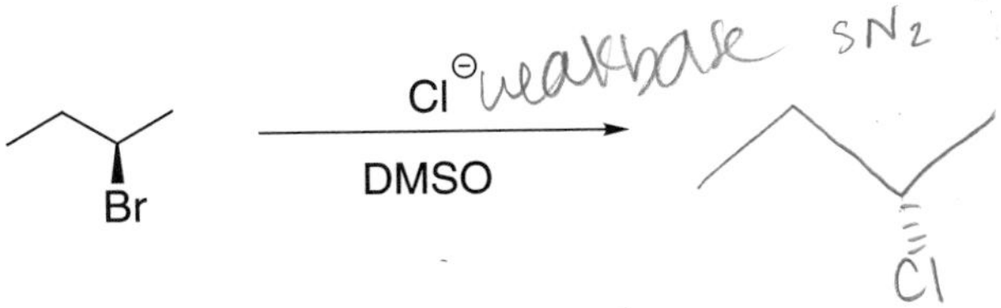

(structure with Br) $\xrightarrow[\text{DMSO}]{\text{Cl}^{\ominus}}$ weak base SN2

12.8

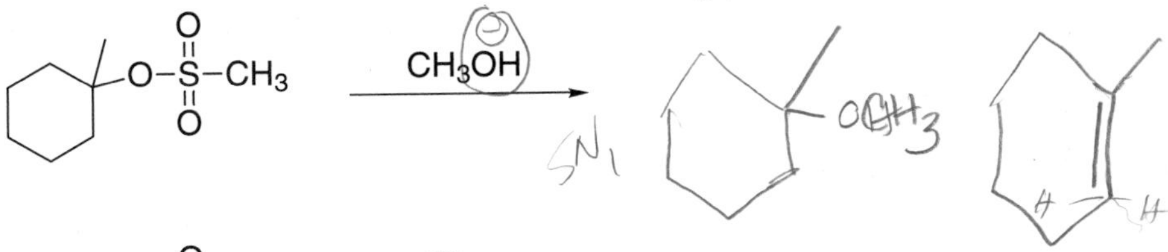

(cyclohexane with O-S(=O)(=O)-CH₃) $\xrightarrow{\text{CH}_3\text{OH}}$ SN1

OCH₃

12.9

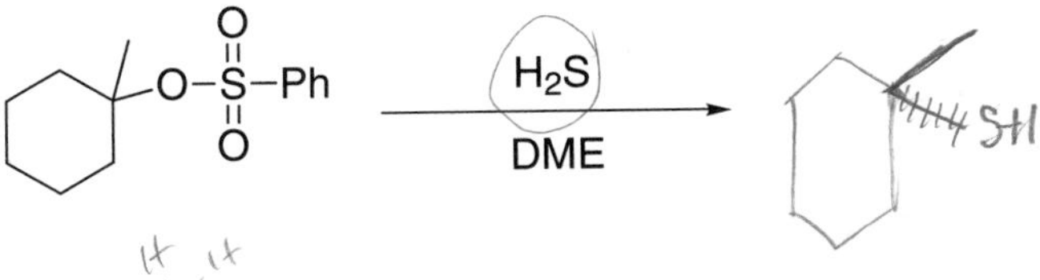

(cyclohexane with O-S(=O)(=O)-Ph) $\xrightarrow[\text{DME}]{\text{H}_2\text{S}}$

SH

12.10

(structure with OH) $\xrightarrow[\text{heat}]{\text{H}_2\text{SO}_4}$

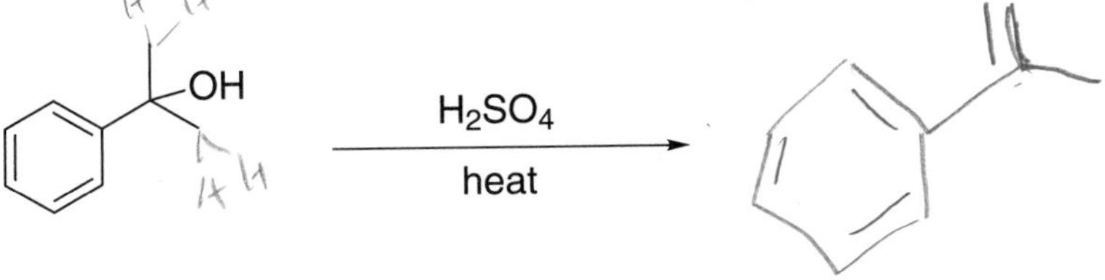

E2

12.11

Hint: You will need to draw a Newman projection for this one.

12.12

12.4 LOOKING FORWARD

You must train yourself to ask the following three questions:

1. What kind of reaction is taking place?
2. What is the regiochemistry of that reaction?
3. What is the stereochemistry of that reaction?

These questions will help you to predict products for all of the reactions that you will see for the rest of the course. Remember that there are three critical questions to ask: If you keep this in mind as you go through the course, you will find that predicting products is not so difficult after all.

CHAPTER 13

SYNTHESIS

Synthesis is really just the flipside of predicting products. In any reaction, there are three groups of chemicals involved: the starting material, the reagents, and the products:

$$\text{Starting material} \xrightarrow{\text{Reagents}} \text{Products}$$

When the products are not shown, then you have a "predict the product" problem:

$$\text{Starting material} \xrightarrow{\text{Reagents}} \text{?}$$

When the reagents are not shown, then you have a synthesis problem:

$$\text{Starting material} \xrightarrow{\text{?}} \text{Products}$$

Now that we see the similarity between predicting products and synthesis, we realize that for every reaction, we need to know the same information that we needed to know for predicting products:

1. What kind of reaction?
2. What is the regiochemistry?
3. What is the stereochemistry?

Remember that all three pieces of information are contained in the mechanism. So your starting point should always be mastery of the mechanisms, followed by a strong understanding of all three pieces of information for every reaction. By doing so, you will have the fundamental building blocks that you need to begin thinking about synthesis problems.

Synthesis problems can be easy (if they are only one step) or they can be difficult (if they are more than one step). When you begin learning reactions in your course, you will start to encounter synthesis problems in your textbook. At first, you will get one-step problems, and as the course progresses, you will see multistep syntheses. In a multistep synthesis, you can often end up with a product that looks very different from the starting material. For example, look at the following series of reactions below. Don't concentrate on how the changes were made. For now, just focus

263

on the fact that each reaction changes the compound only slightly, but in the end, we end up with a product completely different from the starting material:

Starting material

Product

It can only take three or four steps before the problem can get quite difficult. If you convert the sequence above into a synthesis problem, it would look like this:

If you are having trouble with synthesis problems when you first encounter them, the worst thing you can do is to give up and say: "Oh, well, I'm not good at synthesis problems." As the course moves on, this attitude will slowly kill your grade in the course. To see why this is so, let's compare organic chemistry to a game of chess.

Imagine that you are learning how to play chess. You first learn about the pieces: how they are named, how to set up the board, and so on. Then you learn how each piece moves and how they capture each other. When you start playing your first game, you realize that there is quite a bit of strategy involved. Most strategies involve thinking more than just one move in advance. It is not good enough to know only how to move the pieces. You also need to think about how to plan out the next few moves so that you can coordinate an attack on your opponent's pieces. Imagine how silly it would be to take the time to learn how to move the pieces, but to then say to yourself that you are not good at strategy. Imagine thinking that you will keep playing chess, but you just won't be good at that one aspect of the game. That would be silly, because that one aspect of the game is the whole game itself. You either need to learn how to strategize, or just don't play chess. There is no in-between.

Organic chemistry is very much the same. Synthesis is all about strategizing. You need to think a few moves ahead, and you must learn how to do this. You cannot tell yourself that you are not good at synthesis problems, and therefore you will just focus on the other aspects of organic chemistry. Synthesis *is* organic chemistry. The second half of the course is all about learning reactions and applying them in syntheses. Everything that you have learned so far has prepared you for synthesis. The only way to become proficient at synthesis is to *practice*. Don't be lazy, and don't think that you can get through the course without learning how to propose syntheses. If you do, you will find that your performance in the course will spiral down to a point that will make you very unhappy.

There are a few techniques that will make you feel more comfortable with synthesis problems, and there are exercises that you can go through to increase your proficiency in doing synthesis problems. That's what this chapter is all about.

13.1   ONE-STEP SYNTHESES

As we mentioned earlier, one-step syntheses are the first synthesis problems you will encounter. They will never be more difficult than predicting products. Before you can move on to multistep syntheses, you first need to feel comfortable with one-step syntheses.

To do this, we need to make a list, very similar to the one we made in the previous chapter on predicting products. In the list we made last chapter, we left out the products, so that we could repeatedly photocopy the list and fill in the products. This time, we will make a list of the same reactions, but we will leave out the reagents, so that we can repeatedly photocopy the list and get practice filling in the reagents.

As you learn more and more reactions, this list will grow. With every five new reactions, you should photocopy all of the reactions that you have recorded here. Then, start filling in the reagents on the photocopy. If you cannot fill them all in, go back to the previous chapter where you recorded the reactions. Repeat this procedure whenever you have entered five new reactions.

If you keep up with this exercise as the course progresses, you will be in very good shape for solving one-step synthesis problems. The hardest challenge that you will face is keeping up with the work and not waiting until the night before the exam. If you wait (as most students do), you will find it very difficult to spend the time that it takes to master this material. Don't make that mistake. The secret to success in this course is to do a little bit every night (rather than cramming on the night before the exam). Cramming might work well for other courses, but it doesn't work well in organic chemistry.

Begin your list on the next page.

For now, skip forward a few pages. We have some techniques to go over that will help you solve synthesis problems.

Remember not to fill in the reagents or the mechanisms. For each reaction, just draw the starting material in front of the arrow and the products after the arrow. Leave the space above the arrow empty. You will fill in the reagents when you photocopy these pages:

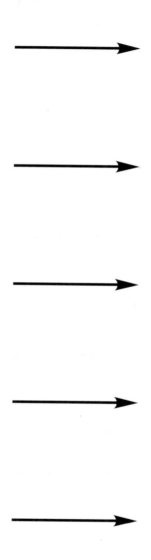

Now photocopy this page, and try to fill in the reagents on your photocopied page.

Now photocopy this page again, and fill in the reagents for every reaction on this page.

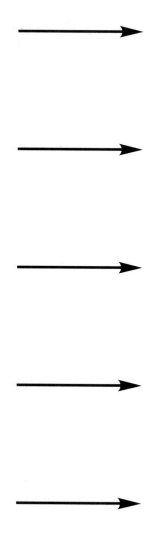

Now photocopy this page AND the previous pages, and fill in all of the reagents.

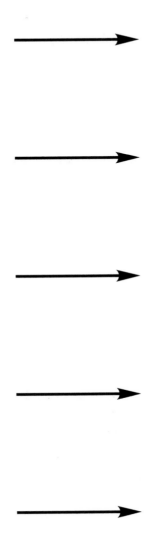

Now photocopy this page AND the previous pages, and fill in all of the reagents.

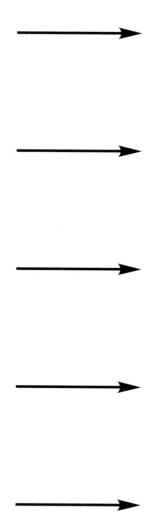

Now photocopy this page AND the previous pages, and fill in all of the reagents.

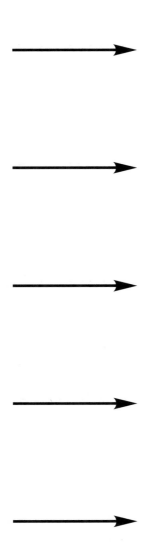

Now photocopy this page AND the previous pages, and fill in all of the reagents.

Now photocopy this page AND the previous pages, and fill in all of the reagents.

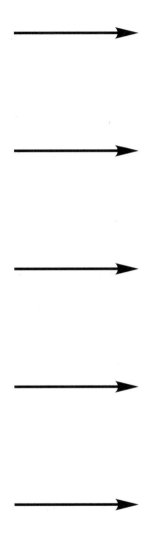

Now photocopy this page AND the previous pages, and fill in all of the reagents.

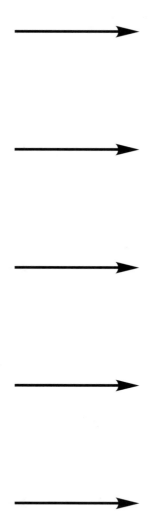

Now photocopy this page AND the previous pages, and fill in all of the reagents.

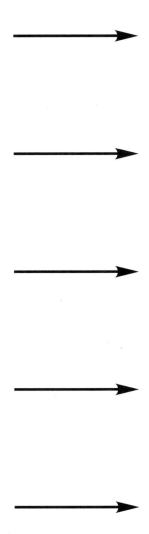

Now photocopy this page AND the previous pages, and fill in all of the reagents.

If you cover more than 30 reactions and need more space to continue, then you can just use a regular piece of paper to keep your list going.

13.2 MULTISTEP SYNTHESES

To prepare yourself for solving multistep syntheses, you need to learn how to think in more than one move. If you carefully review your list of reactions, you will find that the products of some reactions are the starting material for other reactions. For example, you will find that some reactions are used to form double bonds, and other reactions add reagents across double bonds. So if you pair up all of the possibilities, you will create a list of many two-step syntheses. By studying these two-step possibilities, you will begin to get familiar with seeing syntheses that are more than one step.

Let's see an example of what we mean. Below is one reaction that forms a double bond. It starts with an alkyne, and you will certainly learn this reaction at some point:

Now consider one of the reactions where reagents react with a double bond:

If we put these two reactions together into a two-step synthesis, we get the following:

You should now get some practice with this. You will probably learn around five methods for making double bonds and probably around 10 reactions that involve reagents reacting with double bonds. If you put together all of the possibilities, you will find that there are around 50 possibilities, depending on exactly how many reactions you learn. Clearly, you cannot keep a list like this as you go through the course. The list would be too long to study. And if you try to consider three-step syntheses, you will find that the number of permutations is too large to even compile such a list. It's just like our analogy to a game of chess.

In chess, you cannot possibly memorize every possible orientation of all of the pieces and then memorize the best move for each of those possibilities. There are too many permutations. Instead, you learn how to analyze each situation and as time goes on you get better and better at it. By familiarizing yourself with certain permutations, you will get better at figuring things out as you go along. So let's start with the list that we talked about above—the approximately 50 possible two-step syntheses that involve forming a double bond and then doing something to that double bond.

Again, you should not try to make a list like this throughout your entire course. This task would be impractical. But if you make this first list of roughly 50 syntheses, you will learn how to start thinking in more than one step. It is important for you to get accustomed to thinking this way. Take a separate sheet of paper and try to create this list using the reactions in the beginning of your course. If you do not get a chance to write down all 50 reactions, that's OK. As long as you begin the process and draw at least 10 or 20 of them, then you will start to understand what it is like to think in more than one step.

After you have done this, we can start focusing on the main techniques for analyzing problems that display permutations that you have never seen. That is what the next section is all about.

13.3 RETROSYNTHETIC ANALYSIS

When you see a synthesis problem for the first time, you are not expected to immediately know the answer. I cannot stress this enough. It is so common for students to get overly anxious when they see synthesis problems that they cannot solve. Get used to it. This is the way it is supposed to be. Going back to our chess analogy, you don't need to make a move as soon as it is your turn. You are allowed to think about it first. In fact, you are supposed to think about it first. So, how do you begin thinking about a multistep synthesis problem where you do not immediately see the solution? The most powerful technique is called *retrosynthetic analysis*. This means that you analyze the problem backward. Let's see how this works with an example:

The synthesis problem above is a multistep synthesis problem, because we do not have a single reaction that allows us to do this transformation in just one step. So the best way to start is to first look at the product and work our way backward.

We see that the product is a dibromide. So we ask ourselves: Do we know any way of making a dibromide? You can see that to answer this question, you must have first mastered one-step syntheses. If you have not yet done this for all of the reactions that you have learned so far, you will need to go back to the beginning of this chapter and do that first (if you are a student in this situation, continue reading for now, so you can see where this is all going).

So we should be able to recognize that we know how to make dibromides from double bonds. We draw the alkene that would have been used to form the product:

Now we are one step closer to solving this problem. The next step is to ask if there is a way to turn the starting material into this double bond. And there is. We just do an elimination reaction to get the double bond. So now we have solved our synthesis by working backward:

Notice that the stereochemistry and regiochemistry needs to work out for every step. You cannot use a step that has the wrong stereochemistry or regiochemistry. I suppose you could have memorized all possible two-step syntheses from the reactions in your textbook, and then you would have gotten this problem right away (maybe . . .), but that is not a practical approach. What will you do for three-step or four-step syntheses? You need to get accustomed to thinking backward. The more practice you can get, the better off you will be.

Here is where we run into a big problem. There is no way for me to give you problems that are appropriate. Every course goes at its own pace, in its own order, and with exams at different points in the course. I cannot give problems that will be perfectly appropriate for every student everywhere. So, how are we going to get practice? Very simply. You are going to make your own problems. We see how to do this in the next section.

13.4 CREATING YOUR OWN PROBLEMS

Creating your own problems is easier than it sounds. You just choose any reaction from the lists that you have been making (in Chapters 8, 12, and 13). Then look at the product of that reaction and choose another reaction that you have learned that will transform that compound into something else. We work backward to solve the synthesis problems, and we work forward to make up the problems. At each step, draw the product of that reaction and then move on to the next step. After you have gone two or three or four steps, erase everything in the middle. Just draw the very first compound and the final product. Draw an arrow between them, and you have a synthesis problem.

There is one catch. You will not find that problem to be very challenging, because you are the one who made it. So here is what you should do. Find a friend in the course, and each of you should make up 10 or 20 problems. Then you switch off with each other. You will find that this is a very effective method for studying. The larger your study group becomes, the more effective it will be. Don't be shy. You will need to work with a friend to get the practice that you need, not to mention the valuable peer support. If you are reading this book, then chances are that other students in your course have this book also. They will have the same need that you do. Team up with them.

Even if you cannot find a friend with whom you can swap problems, it will still be a useful exercise to create your own problems. The process of creating problems by itself is a worthwhile process. It will help you get accustomed to thinking in multiple steps for synthesis problems.

To summarize, these are the keys to becoming proficient at solving synthesis problems:

1. Master the one-step syntheses by constant review.

2. Train yourself to work backward when solving a problem.

3. And, finally, get lots of practice.

ANSWER KEY

Chapter 1

1.2) 12
1.3) 6
1.4) 6
1.5) 6
1.6) 5
1.7) 6
1.8) 7
1.9) 8
1.10) 7
1.11) 4

1.13)

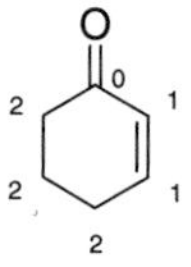

1.14)

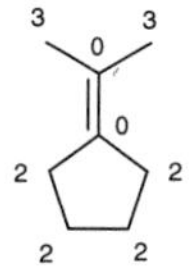

1.15)

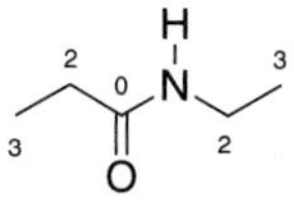

1.16)

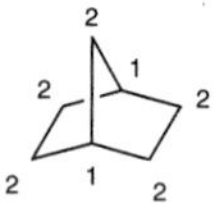

1.17)

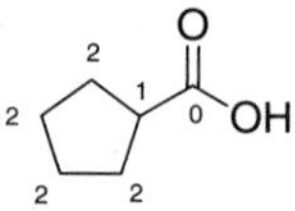

1.18)

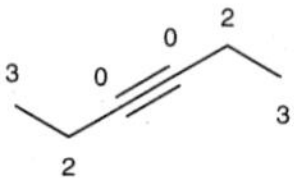

1.19)

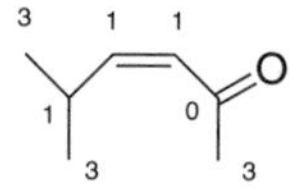

1.20)

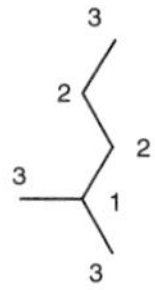

1.21)

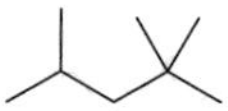

1.22)

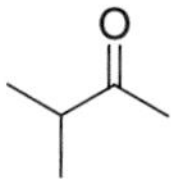

1.23)

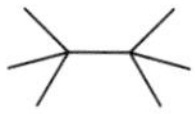

1.24)

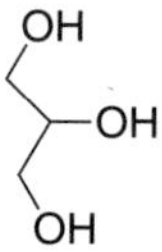

1.25) Substituted a Cl with an OH
1.26) Added two OH groups across a double bond
1.27) Eliminated H and Cl to form a double bond
1.28) Added Br and Br across a double bond
1.29) Eliminated H and H to form a double bond
1.30) Substituted an I with an SH
1.31) Eliminated H and H to form a triple bond
1.32) Added H and H across a triple bond
1.34) No charge
1.35) Positive
1.36) Negative
1.37) No charge
1.38) Positive

280

1.39) Negative
1.40) Positive
1.41) Negative
1.42) No charge
1.43) Positive
1.44) No charge
1.45) No charge

1.47)

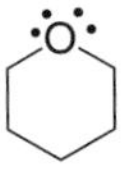

1.48)

1.49)

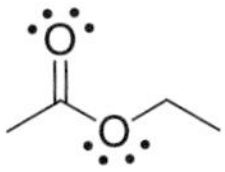

1.50)

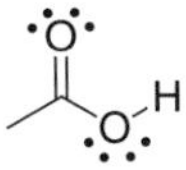

1.51)

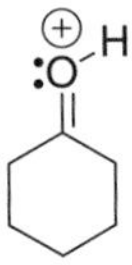

1.52)

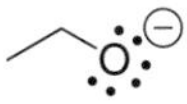

1.54)

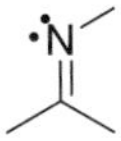

1.55)

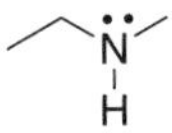

1.56)

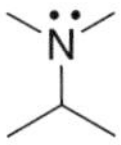

1.57)

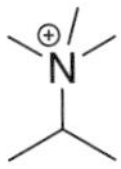

1.58)

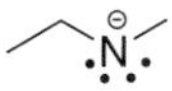

1.59)

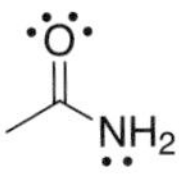

1.60)

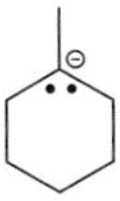

1.61)

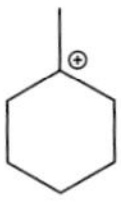

1.62)

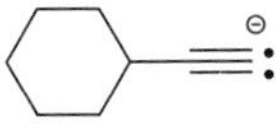

1.63)

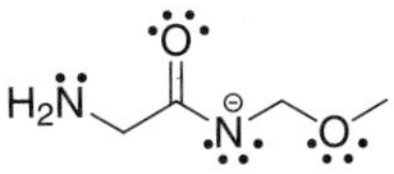

1.64)

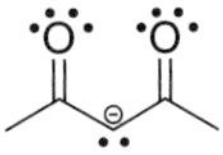

1.65)

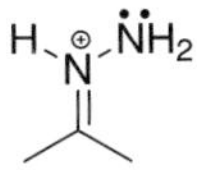

1.66)

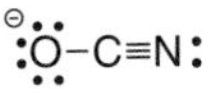

1.67)

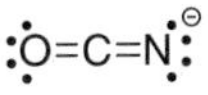

1.68)

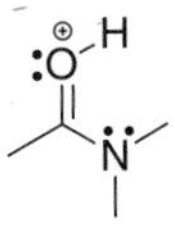

Chapter 2

2.2) Violates second commandment—nitrogen cannot have five bonds
2.3) No violation
2.4) Violates second commandment—carbon cannot have five bonds
2.5) Violates second commandment—oxygen cannot have three bonds and two lone pairs
2.6) Violates second commandment—carbon cannot have five bonds
2.7) Violates second commandment—carbon cannot have five bonds
2.8) Violates both commandments
2.9) Violates both commandments

2.10) No violation

2.11) No violation

2.12) Violates second commandment—
carbon cannot have five bonds

2.14)

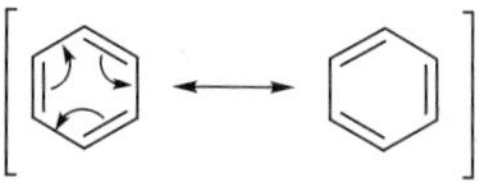

2.15)

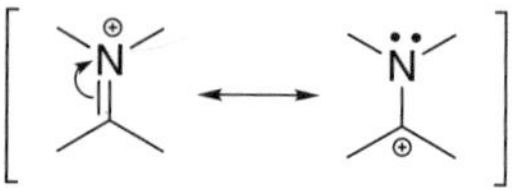

2.16)

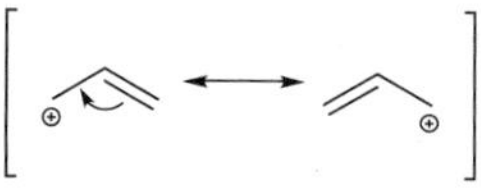

2.17)

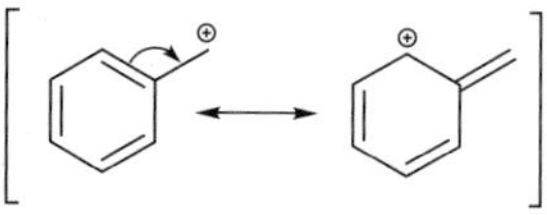

2.18)

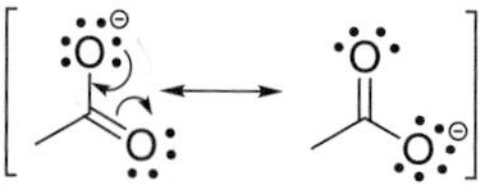

2.19)

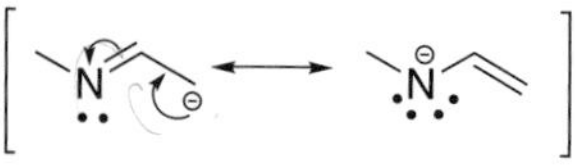

2.21)

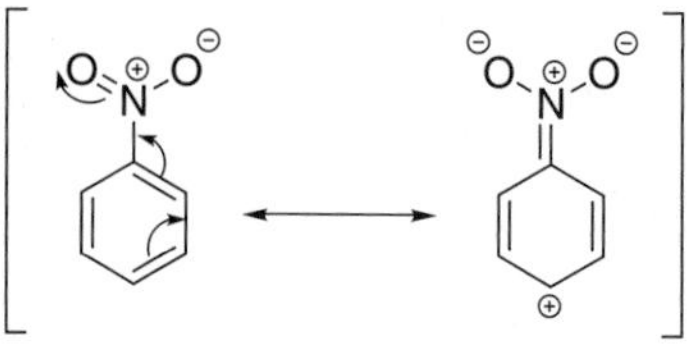

2.22)

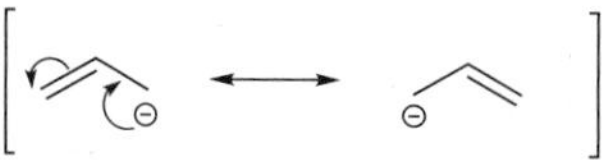

2.23)

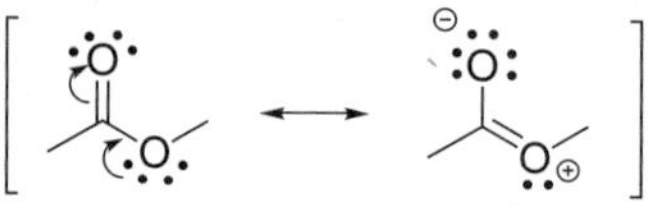

2.24)

2.25)

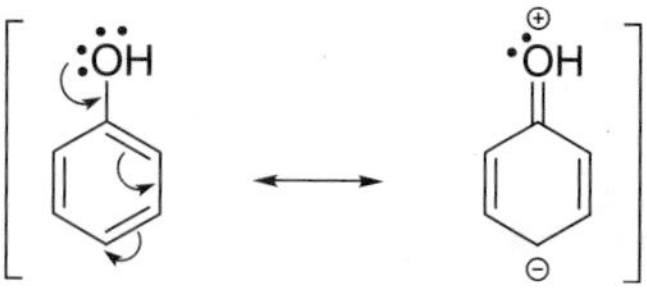

2.26)

2.27)

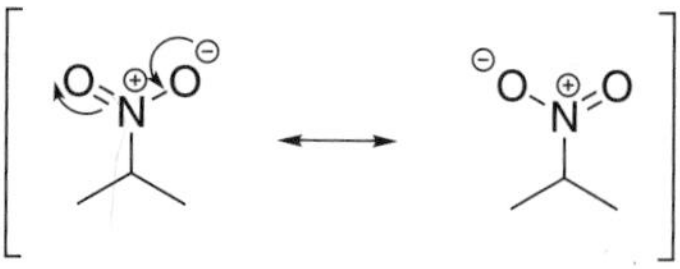

2.28)

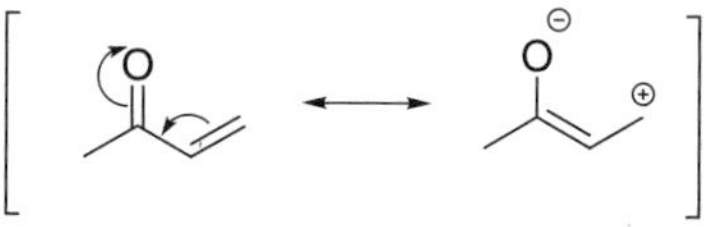

2.30)

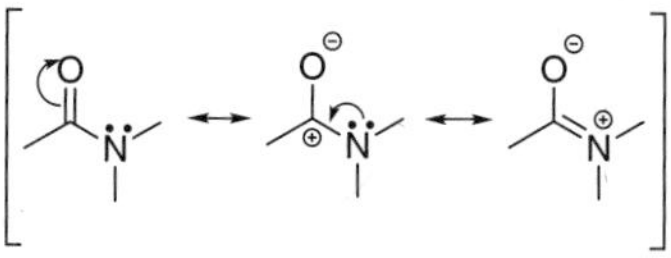

2.32)

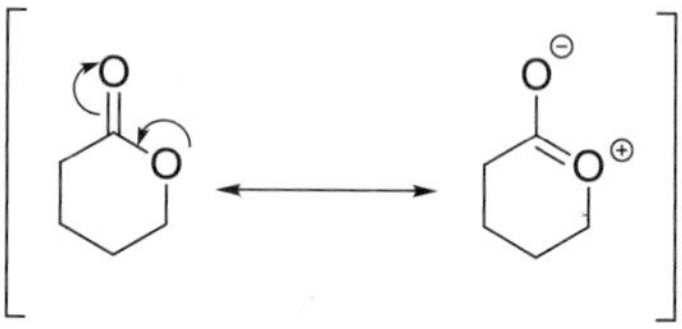

2.33)

2.34)

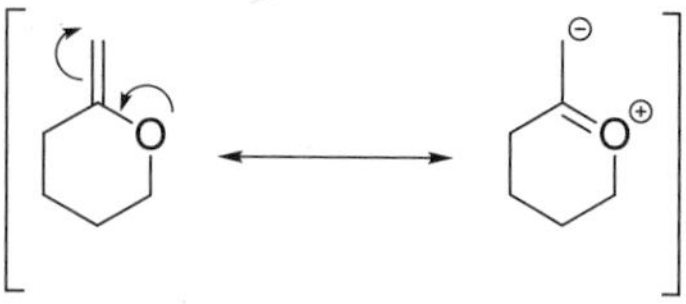

2.35)

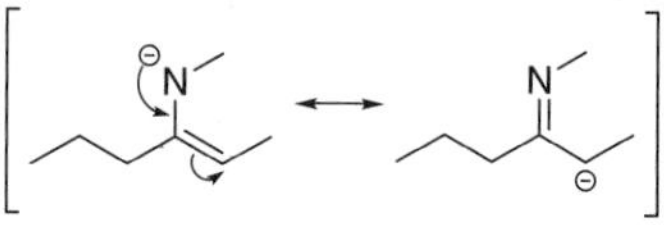

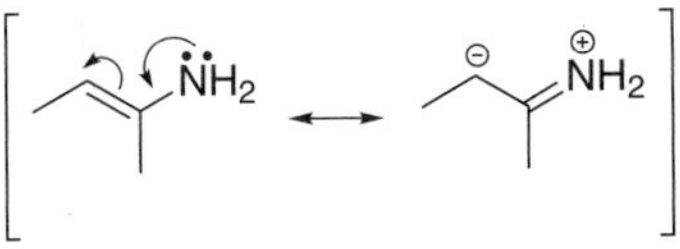

2.36)

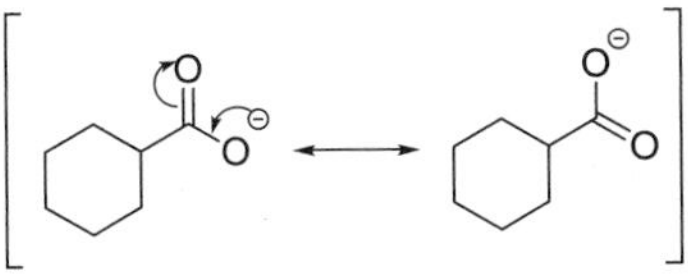

2.37)

2.38)

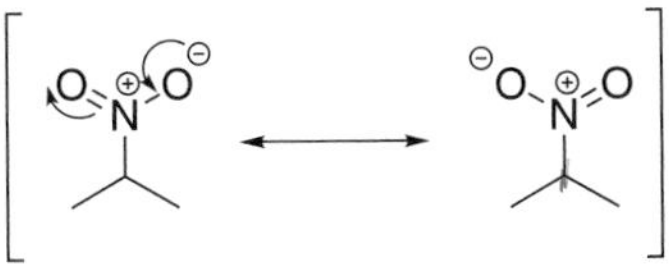

2.39)

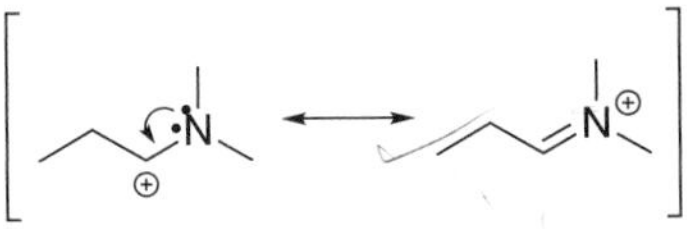

2.40)

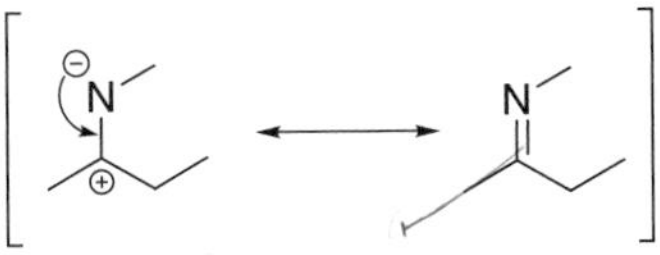

2.41)

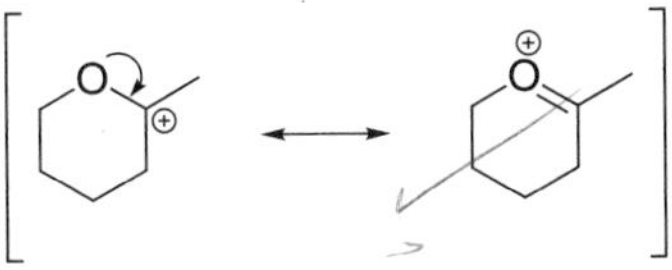

2.42)

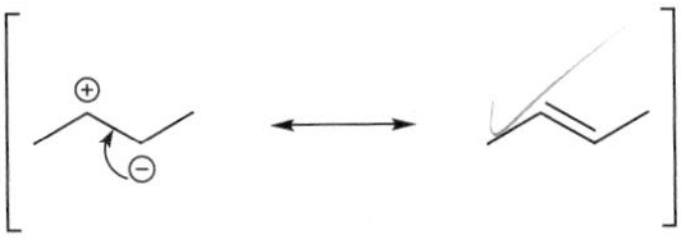

2.43)

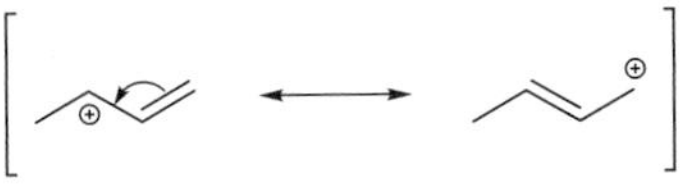

2.44)

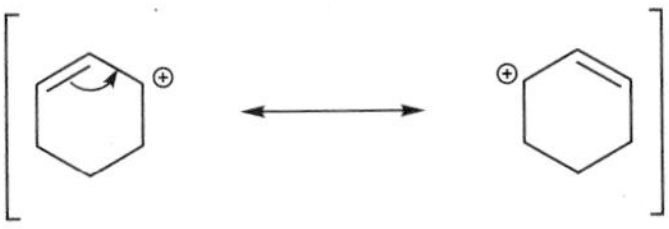

2.45)

2.46)

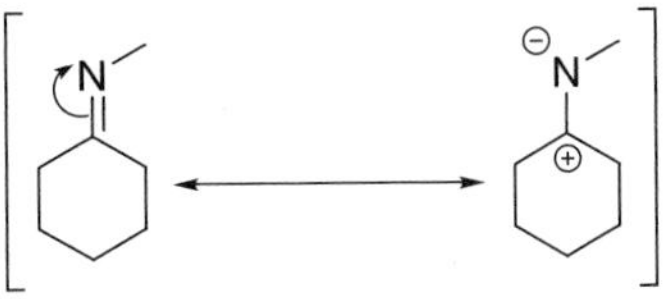

2.47)

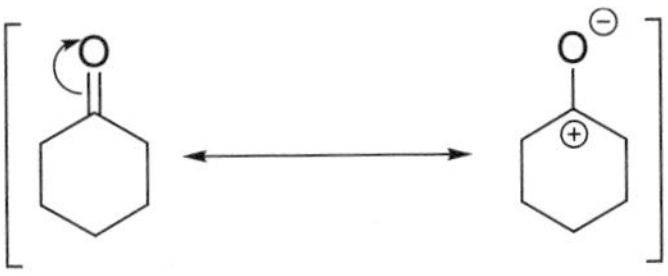

2.48)

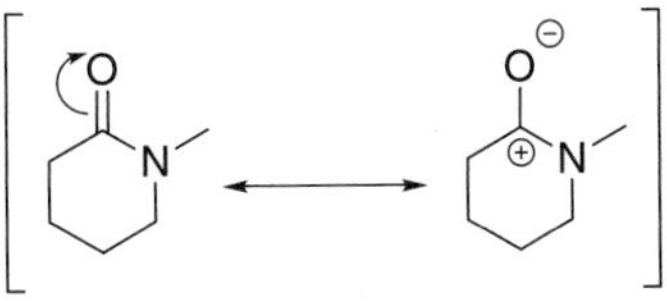

2.49)

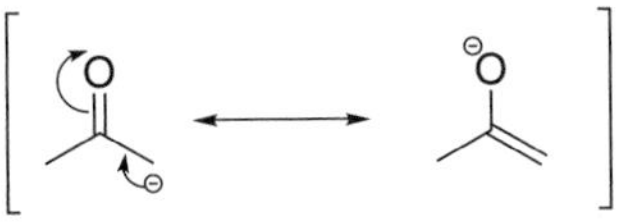

2.50)

2.51)

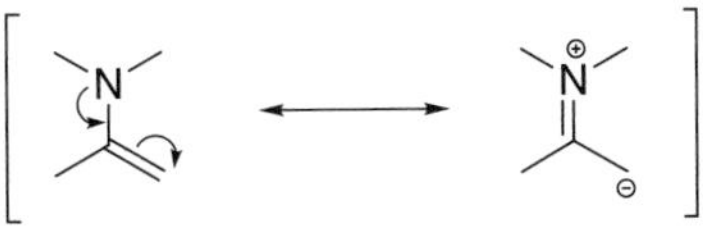

2.52)

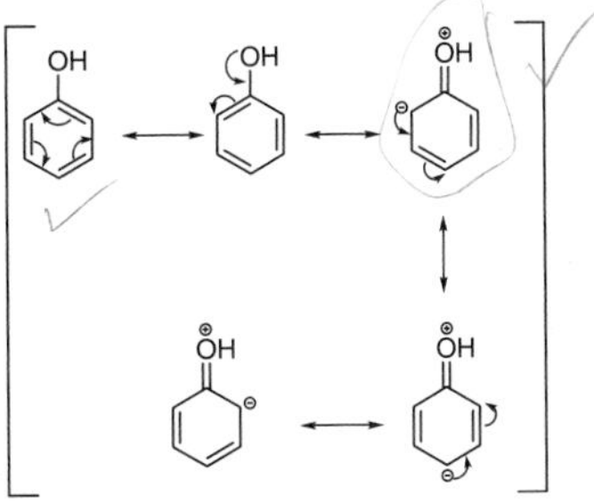

2.53)

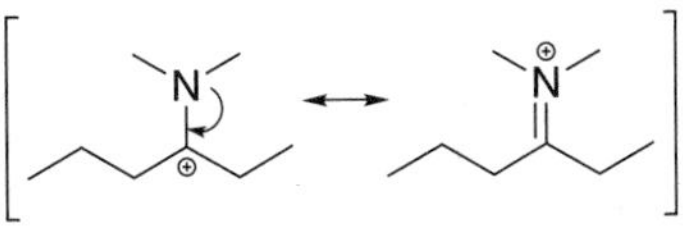

2.54)

2.55)

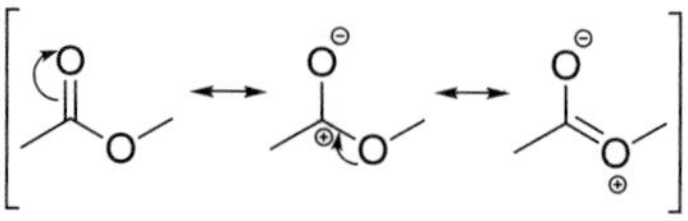

2.56)

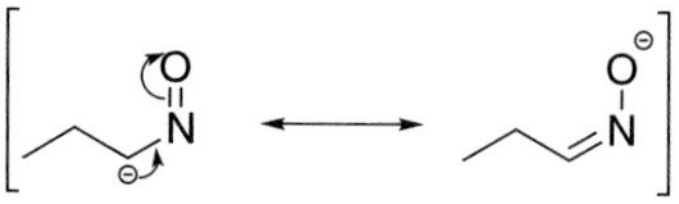

2.57)

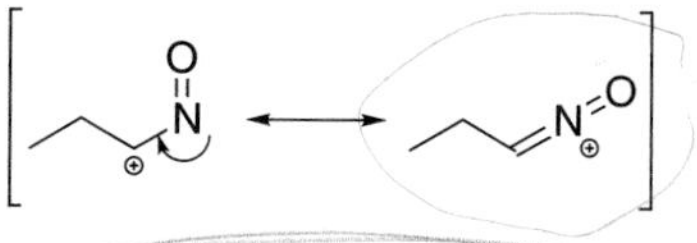

2.58)

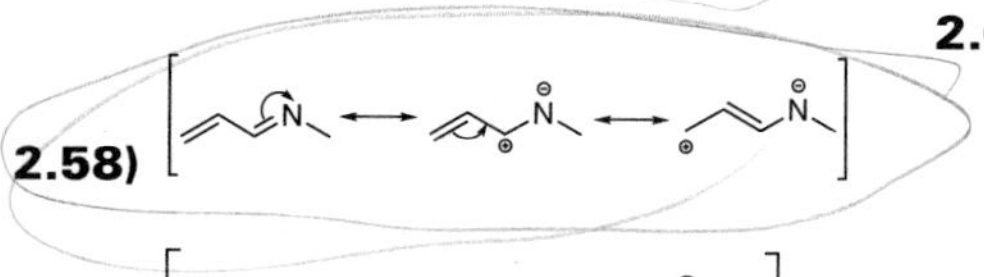

2.59)

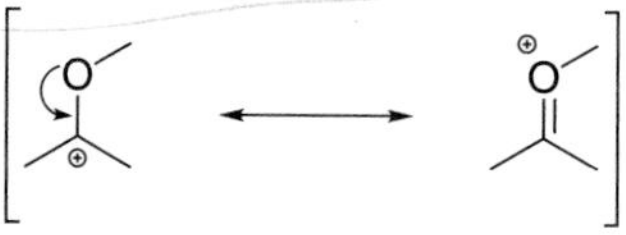

2.60)

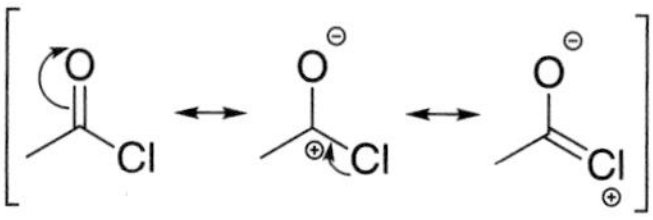

2.61)

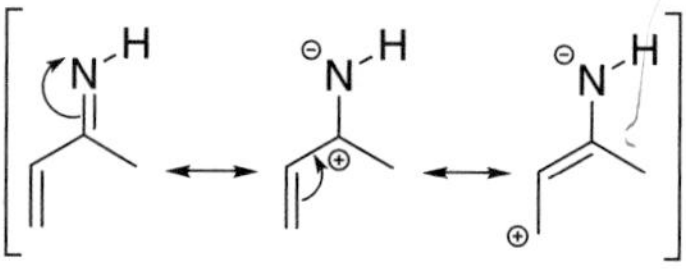

2.62)

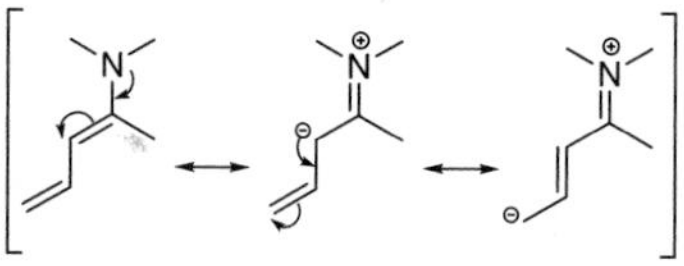

2.63)

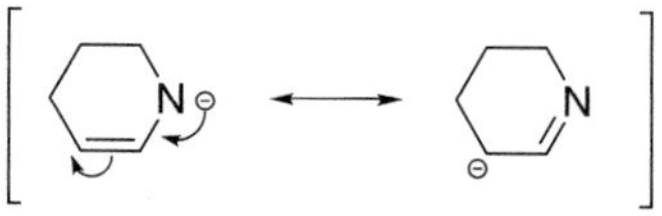

2.64)

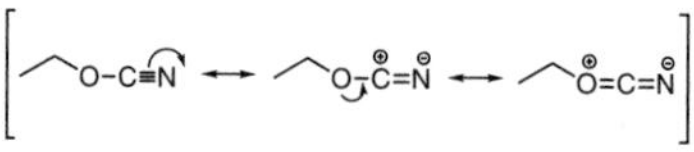

2.65)

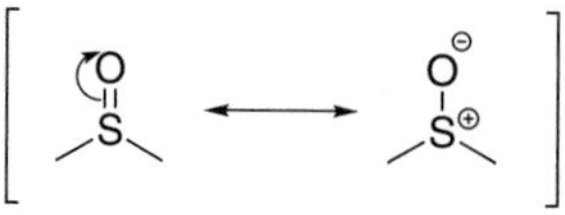

2.66)

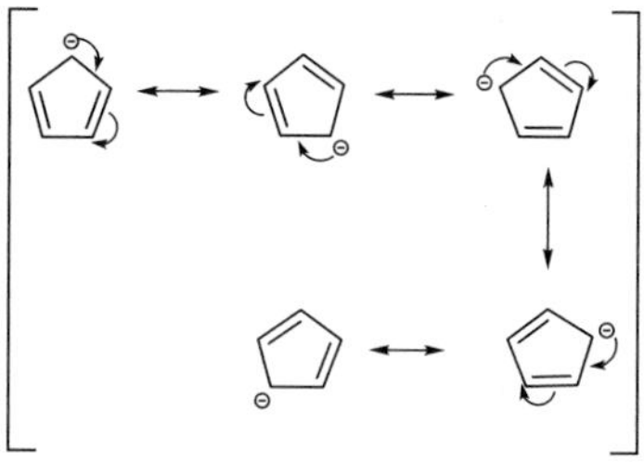

2.67)

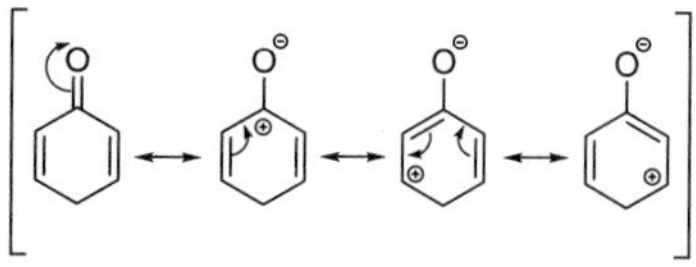

2.68)

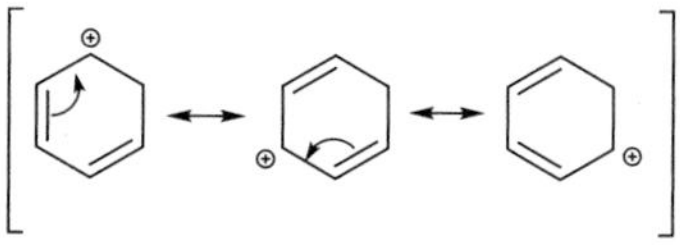

2.69)

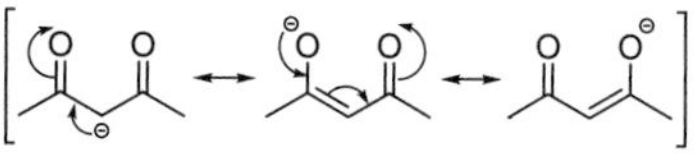

2.70)

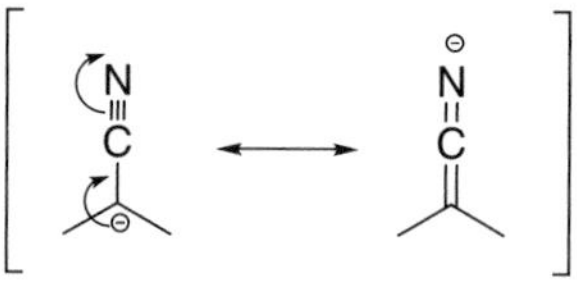

2.71)

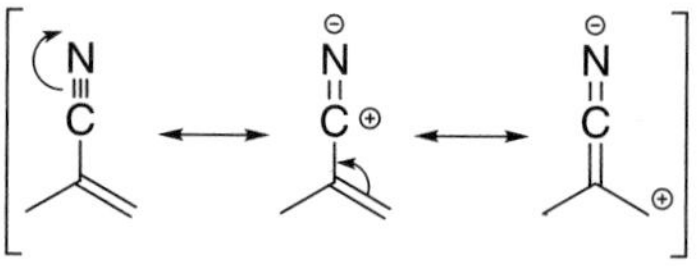

2.72)

2.73)

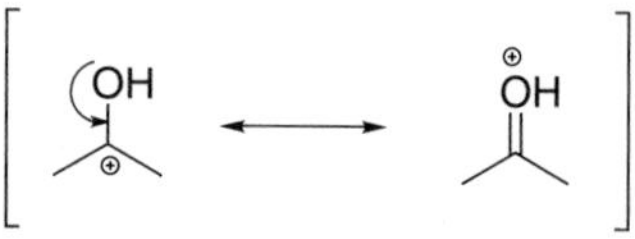

2.74)

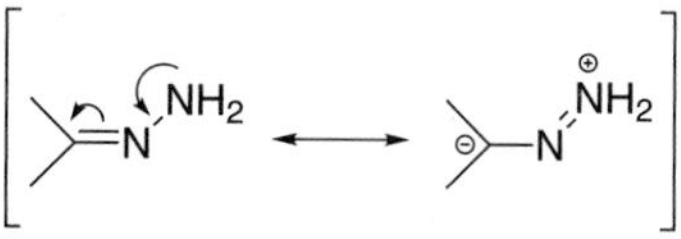

2.75)

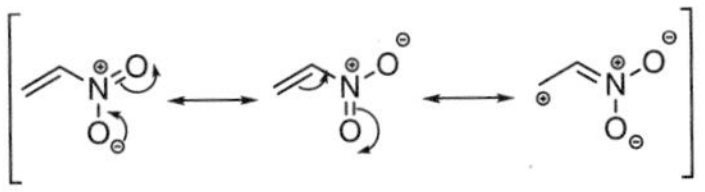

Note: In the last resonance structure, there are more than two charges. Generally, a resonance structure with 4 charges would *not* be significant. But examples with a nitro group are exceptions, because the nitro group already has two charges (positive and negative) associated with it (see page 43). For molecules containing a nitro group, it is acceptable to draw a resonance structure that gives two more charges in addition to the two already there for the nitro group.

Chapter 3

3.2)

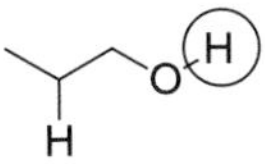

3.3)

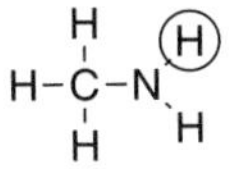

3.4)

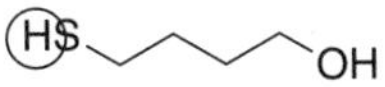

3.5)

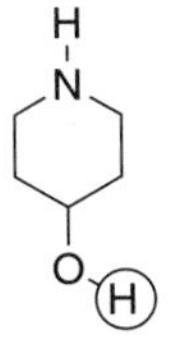

3.7)

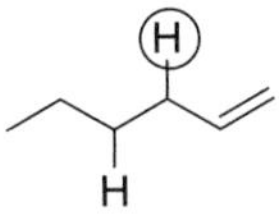

3.8)

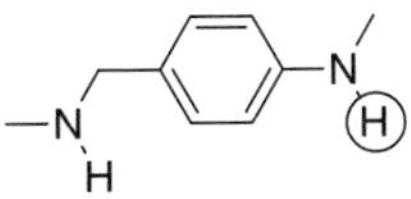

3.9)

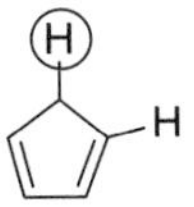

3.10)

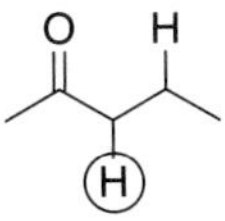

3.11)

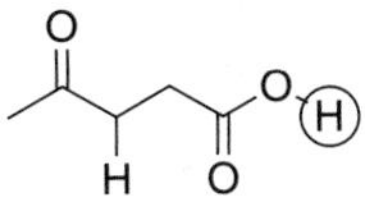

3.12)

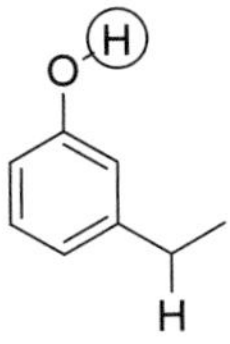

3.14)

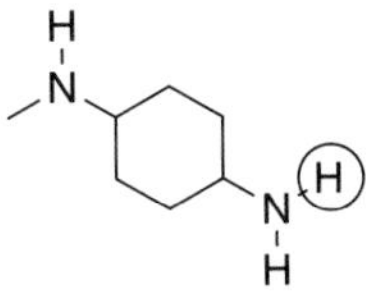

3.15)

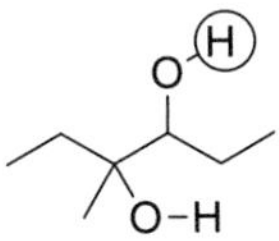

3.16)

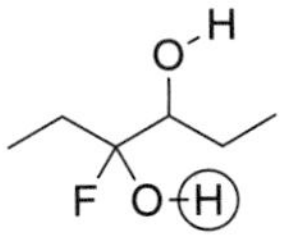

3.19)

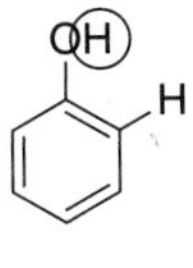

3.20)

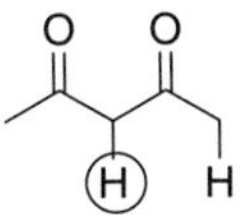

3.21)

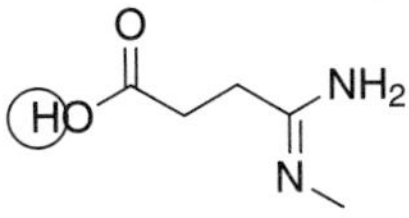

3.22)

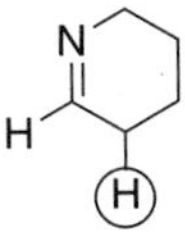

3.23)

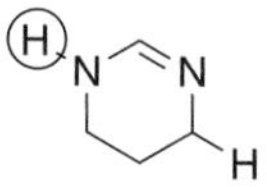

3.24)

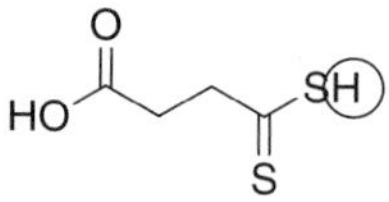

3.25)

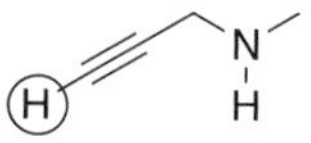

3.26)

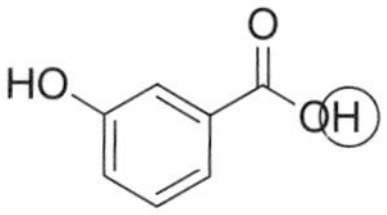

3.27)

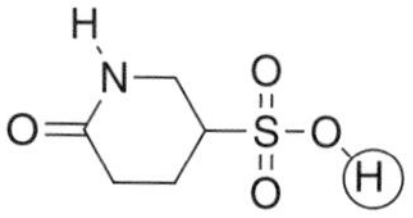

3.28) HBr

3.29) H_2S

3.30) NH_3

3.31) H——≡——H

3.32)

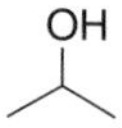

3.33) Cl_3C⟍CCl_3 (OH)

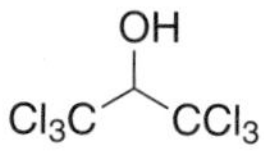

3.35) ⟶

3.36) ⟵

3.37) ⟶

3.39)

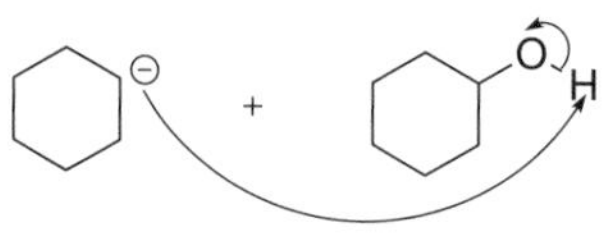

3.40)

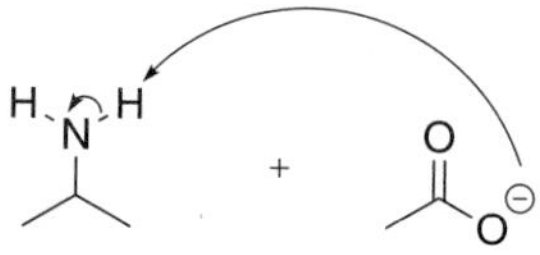

3.41)

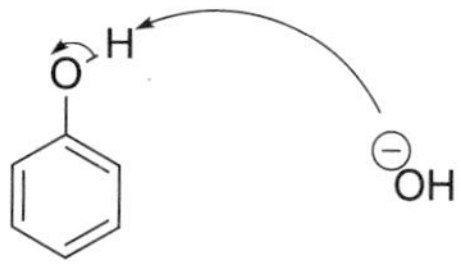

3.42)

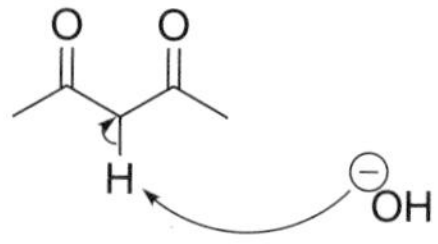

3.43)

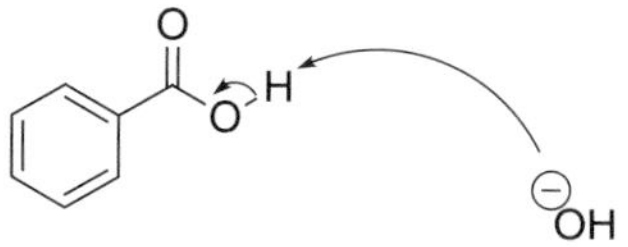

3.44)

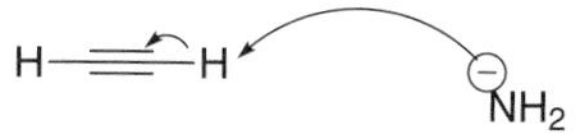

3.45)

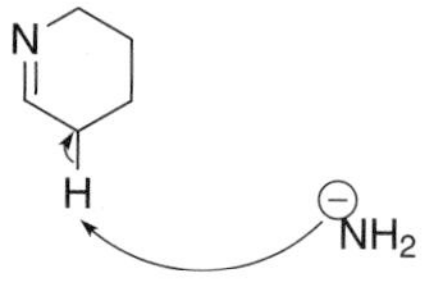

3.46)

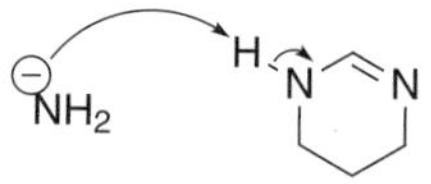

Chapter 4

4.2) sp^2

4.3) sp

4.4) sp^3

4.5) sp^2

4.6) sp^3

4.7) sp

4.8)

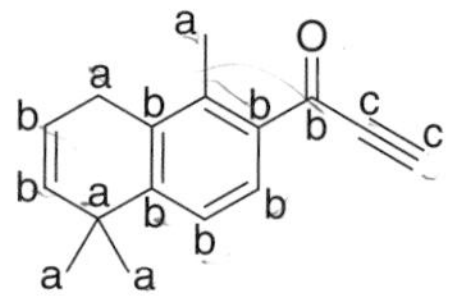

$a = sp^3$

$b = sp^2$

$c = sp$

4.10) All are sp^2 and trigonal planar

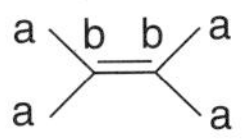

a = tetrahedral, sp^3
4.11) b = trigonal planar, sp^2

a = tetrahedral, sp^3
b = trigonal planar, sp^2 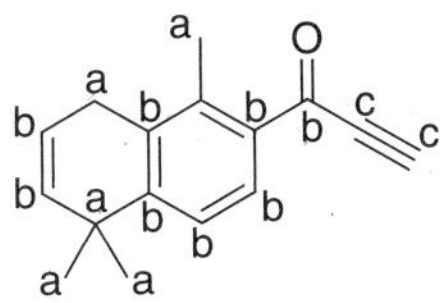
4.12) c = linear, sp

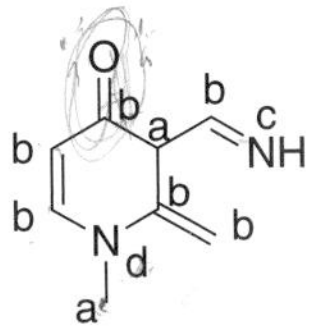

a = tetrahedral, sp^3
b = trigonal planar, sp^2
c = bent, sp^2
4.13) d = trigonal pyramidal, sp^3

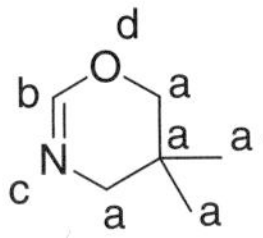

a = tetrahedral, sp^3
b = trigonal planar, sp^2
c = bent, sp^2
4.14) d = bent, sp^3

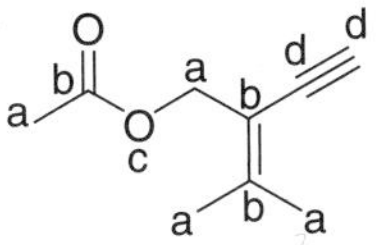

a = tetrahedral, sp^3
b = trigonal planar, sp^2
c = bent, sp^3
4.15) d = linear, sp

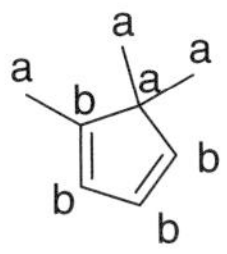

a = tetrahedral, sp^3
4.16) b = trigonal planar, sp^2

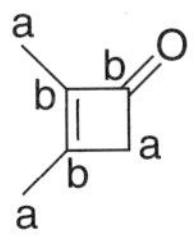

a = tetrahedral, sp^3
4.17) b = trigonal planar, sp^2

Chapter 5

5.2) –one
5.3) –oate
5.4) –al
5.5) –amine
5.6) –ol
5.7) –ol
5.8) –al
5.9) –one
5.10) –oic acid
5.12) –en-
5.13) –yn-
5.14) –dien-
5.15) –trien-
5.16) –trien-
5.17) –endiyn-
5.19) hex
5.20) hept
5.21) hex
5.22) non
5.23) oct
5.24) hex
5.25) hex
5.26) hex
5.27) pent
5.29) Two chloro groups
5.30) bromo, iodo
5.31) Five methyl groups
5.32) Six fluoro groups
5.33) Methyl
5.34) chloro, *tert*-butyl
5.35) amino, bromo, chloro, fluoro

5.36) iodo, fluoro, bromo

5.37) isopropyl

5.38) ethyl, hydoxy

5.40) trans

5.41) trans

5.42) trans

5.43) cis

5.44) cis

5.45) trans

5.48)

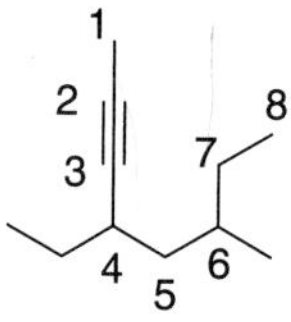

5.49)

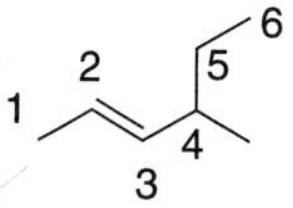

5.50)

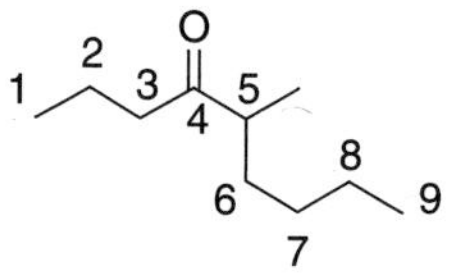

5.51)

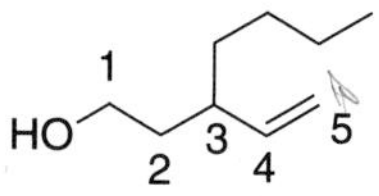

5.52)

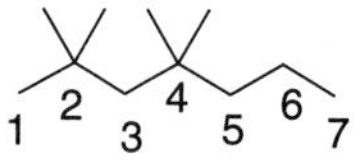

5.53)

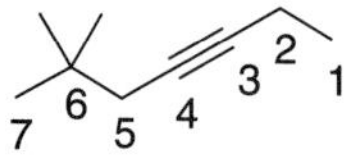

5.54)

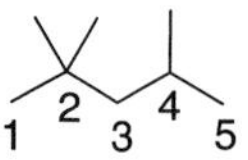

5.55)

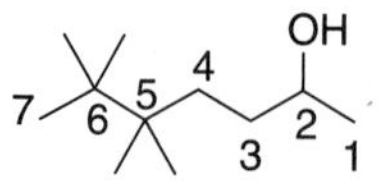

5.56)

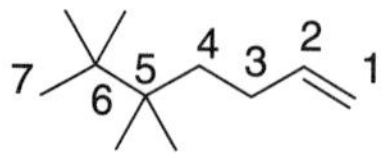

5.57) *trans*-5-ethyl-4-methyloct-2-ene

5.58) 4-ethylnonan-3-ol

5.59) 4,4-dimethylhex-2-yne

5.60) 4,4-dimethylcyclohexanone

5.61) 2-chloro-4-fluoro-3,3-dimethylhexane

5.62) *cis*-3-methylhex-2-ene

5.63) 2-ethylpentanamine

5.64) 2-propylpentanoic acid

5.65) *trans*-oct-2-en-4-ol

5.66) *trans*-5-chloro-6-fluoro-5,6-dimethyloct-2-ene

Chapter 6

6.2)

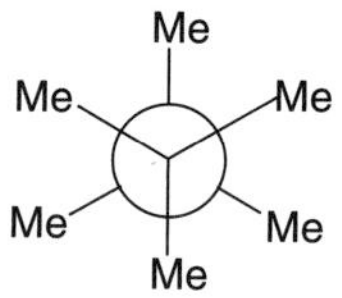

6.3)

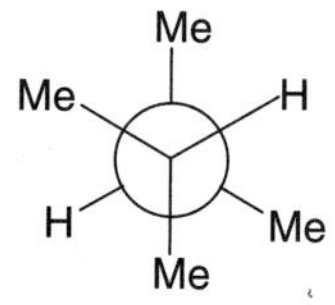

6.4)

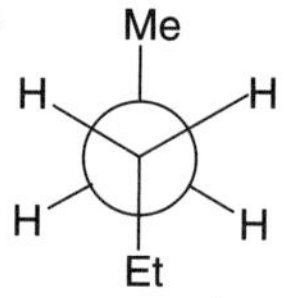

6.5)

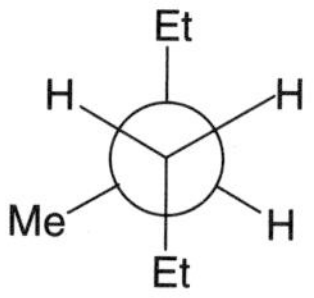

6.6)

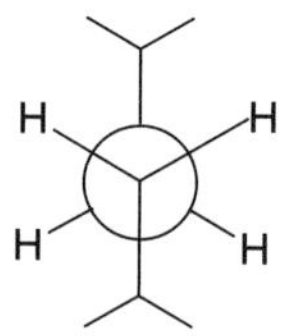

6.7)

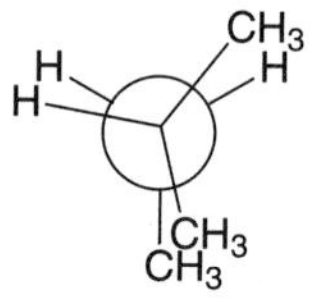

6.9) Me, Me, Me, Me, Me (Newman projection) · H₃C, CH₃, H₃C, CH₃, CH₃, CH₃

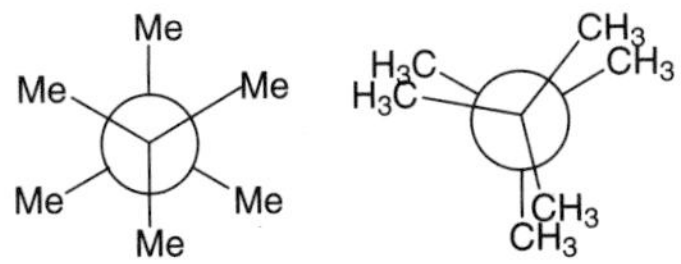

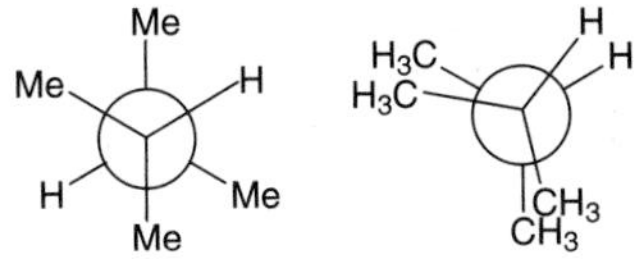

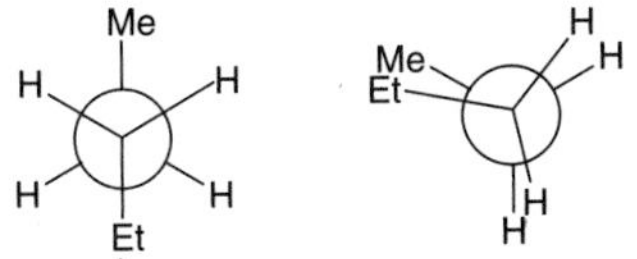

6.10) Me, Me, Me, H, H, Me · H₃C, H₃C, H, H, CH₃, CH₃

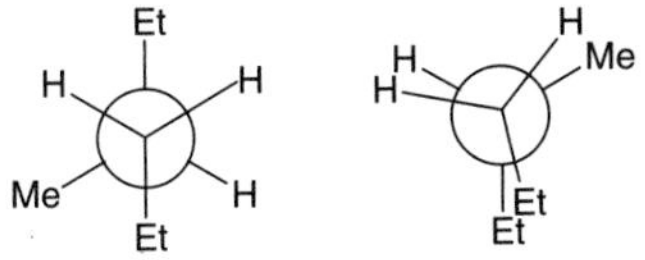

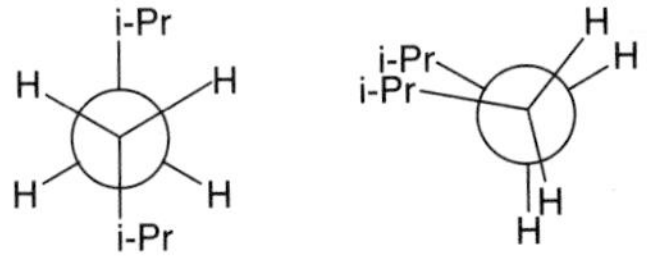

6.11) Me, H, H, H, H, Et · Me, Et, H, H, H

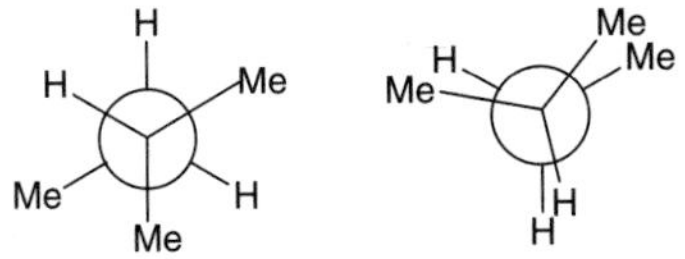

6.12) Et, H, H, Me, Et, H · H, H, H, Me, Et, Et

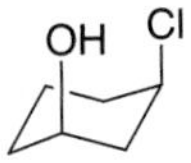

6.13) i-Pr, H, H, H, H, i-Pr · i-Pr, i-Pr, H, H, H

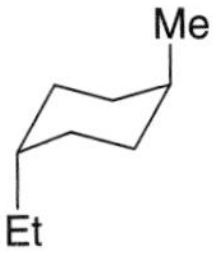

6.14) H, H, Me, Me, H, Me · H, Me, Me, Me, H, H

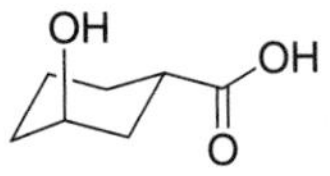

6.16) OH, Cl

6.17) Me, Et

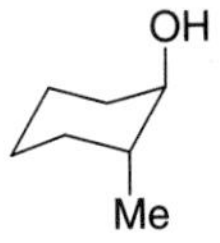

6.18) Me, Br

6.19) OH, OH, O

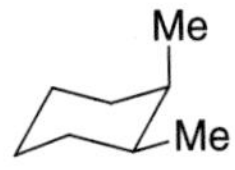

6.20) OH, Me

6.21) Me, Me

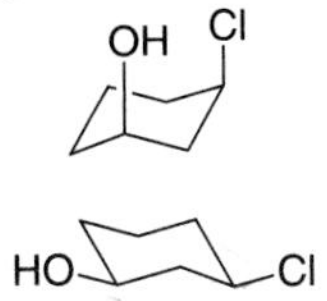

6.24) OH, Cl · HO, Cl

6.25) Me, Et · Et, Me

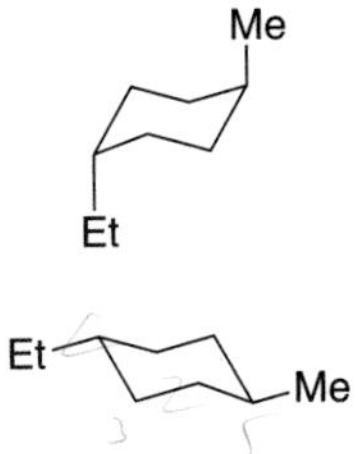

6.26) Br, Me · Me, Br

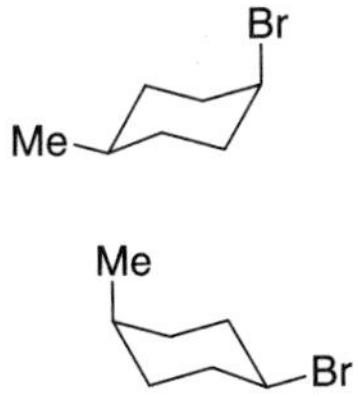

6.27) OH, OH, O · HO, COOH

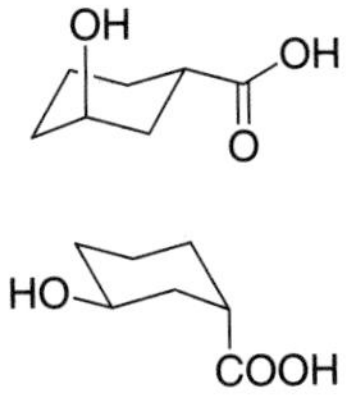

6.28) OH, Me · Me, OH

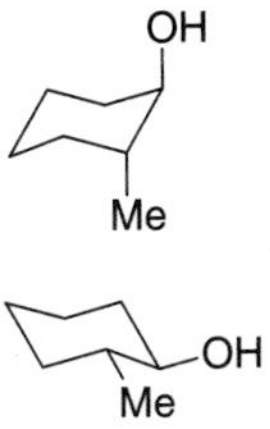

6.29)

6.31)

6.32)

6.33)

6.34)

6.35)

6.36)

6.38)

6.39)

6.40)

6.41)

6.42)

6.43)

6.44)

6.45)

Chapter 7

7.2)

7.3)

7.4)

7.5)

7.6)

7.7)

7.9)

7.10)

7.11)

7.12)

7.13)

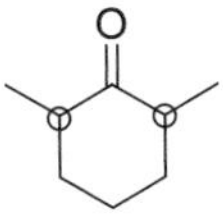

7.14)

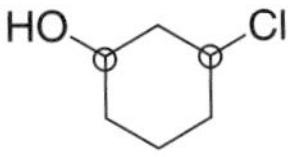

7.15)

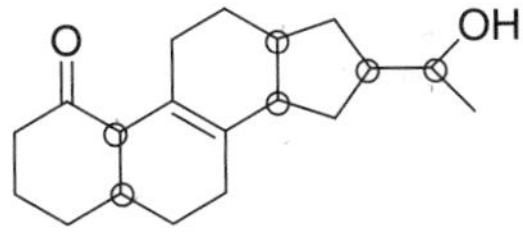

7.17)

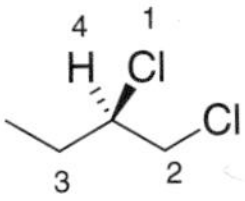

7.18)

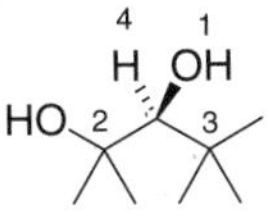

7.19)

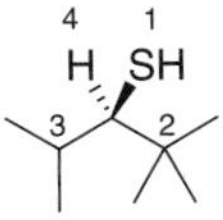

7.21)

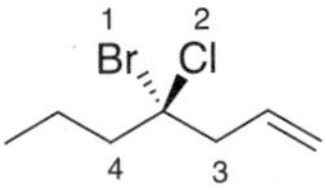

7.22)

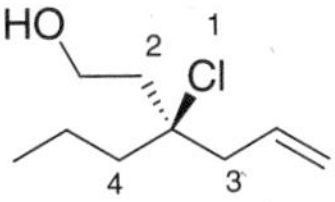

7.23)

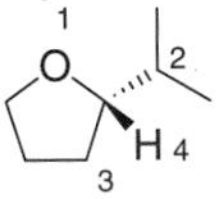

7.24)

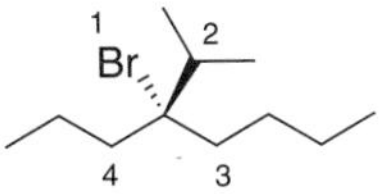

7.25)

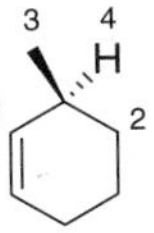

7.26)

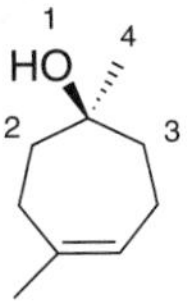

7.27) S
7.28) R
7.29) S
7.30) S
7.31) R
7.32) R
7.33) R
7.34) R
7.35) S

7.37)

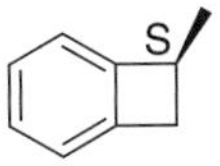

7.38)

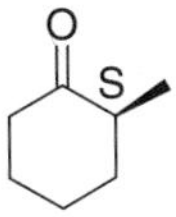

7.39)

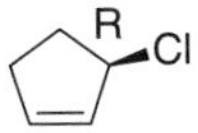

7.40)

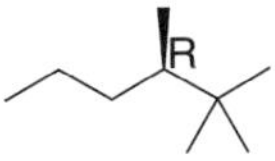

7.41)

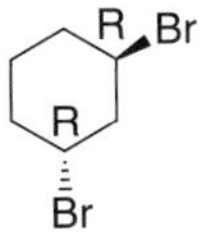

7.42)

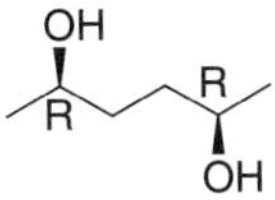

7.44) Z-2-fluoropent-2-ene
7.45) 1R,3R-3-methylcyclohexan-1-ol

7.46) *S*-3-methylpent-1-ene

7.47) *E*-4-ethyl-2,3-dimethylhept-3-ene

7.48) 2*E*,4*Z*-hexa-2,4-diene

7.49) 2*E*,4*Z*,6*Z*,8*E*-deca-2,4,6,8-tetraene

7.62)

7.51)

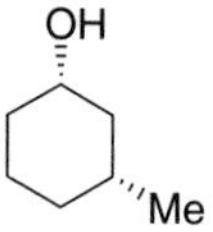

7.52)

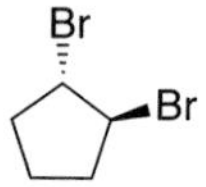

7.53)

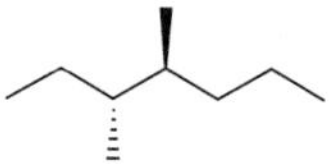

7.54)

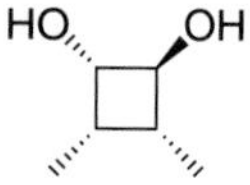

7.55)

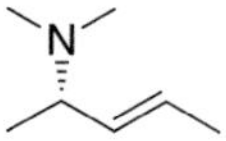

7.56)

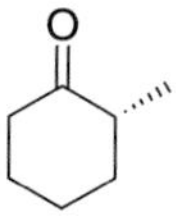

7.58)

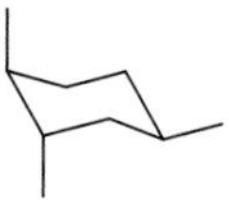

7.59)

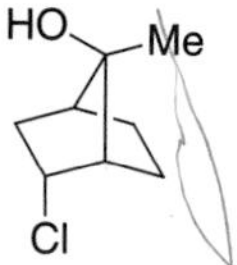

7.60)

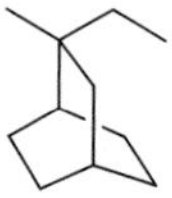

7.61)

7.63)

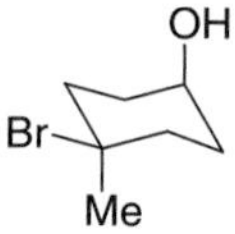

7.65) Enantiomers

7.66) Diastereomers

7.67) Enantiomers

7.68) Enantiomers

7.69) Diastereomers

7.70) Diastereomers

7.72) Meso

7.73) Not meso

7.74) Meso

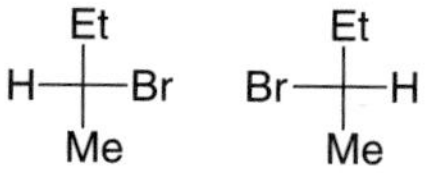

7.76) R

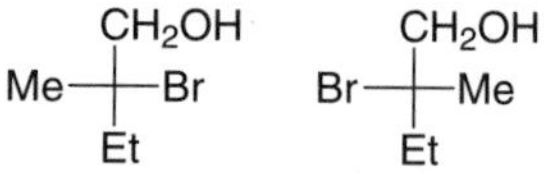

7.77) R

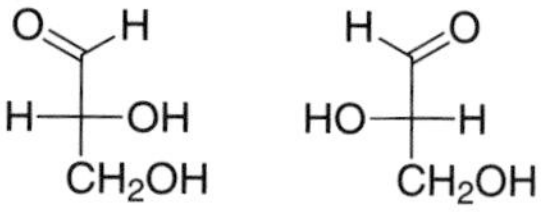

7.78) R

7.79)

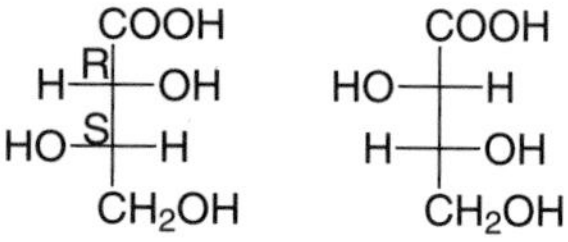

7.80)

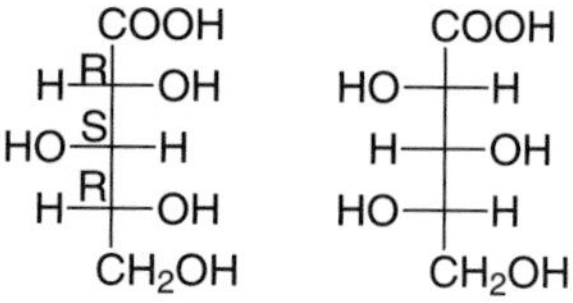

7.81)

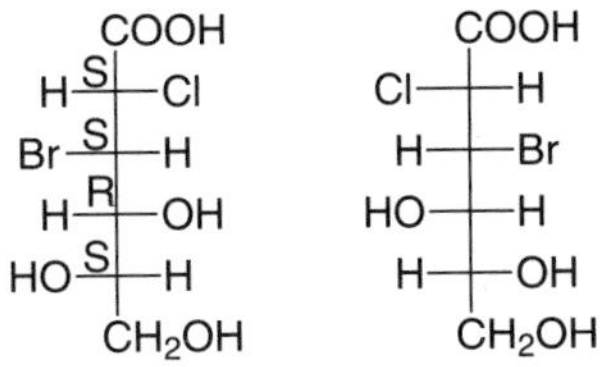

$$\begin{array}{c}
COOH \\
H \overset{S}{-} Cl \\
Br \overset{S}{-} H \\
H \overset{R}{-} OH \\
HO \overset{S}{-} H \\
CH_2OH
\end{array}
\qquad
\begin{array}{c}
COOH \\
Cl - H \\
H - Br \\
HO - H \\
H - OH \\
CH_2OH
\end{array}$$

Chapter 8

8.2) Bond → Lone pair

8.3) Lone pair → Bond, then Bond → Lone pair

8.4) Lone pair → Bond, then Bond → Lone pair

8.5) Lone pair → Bond, then Bond → Lone pair

8.6) Lone pair → Bond, Bond → Bond, Bond → Lone pair

8.7) Lone pair → Bond, then Bond → Lone pair

8.9) Cl⁻

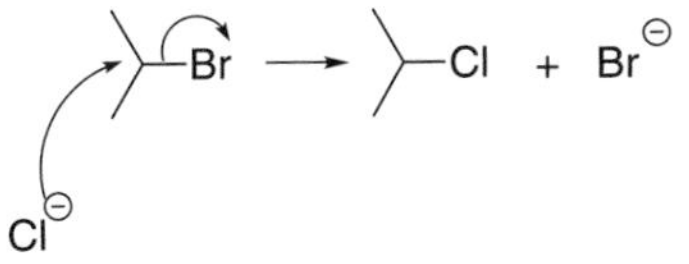

8.10)

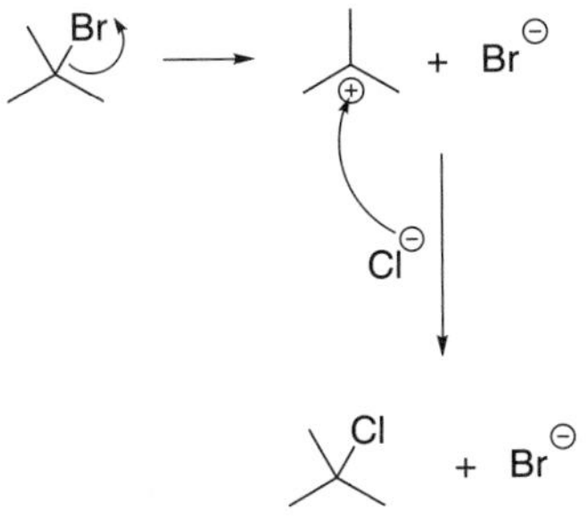

8.11)

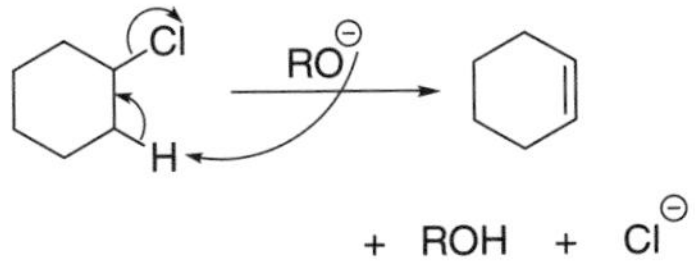

8.12)

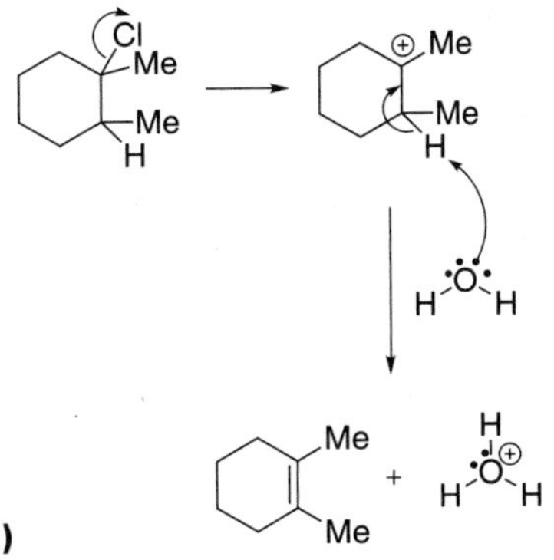

8.14)

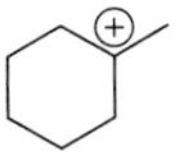

8.15)

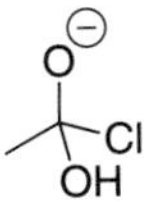

8.16)

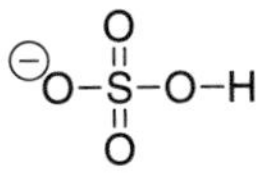

8.17)

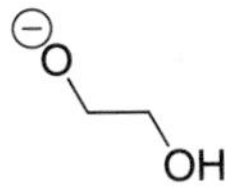

8.18)

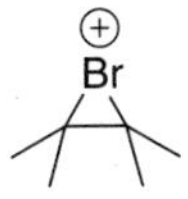

8.19)

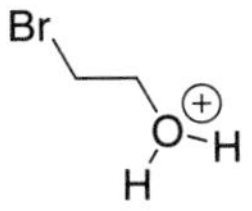

8.21) Hydroxide is the nucleophile

8.22) Water is the nucleophile

8.23) Water is the nucleophile

8.24) MeCl is the electrophile

8.26) Nucleophile

8.27) Base

8.28) Nucleophile

8.29) Base

8.30) Base

8.31) Nucleophile

8.32) Nucleophile

8.33) Base

8.34)

8.35) Br

8.36) trans   cis

8.37)

8.39)

+

8.40)

+

8.41)

8.42)

+

Chapter 9

9.2) Both
9.3) S_N2
9.4) Both
9.5) S_N1
9.7) No
9.8) Yes
9.9) No
9.10) Yes
9.12) Both
9.13) S_N1
9.14) S_N2
9.15) S_N2
9.16) S_N1
9.17) Both

9.19) Good
9.20) Excellent
9.21) Good
9.22) Bad
9.23) Good
9.24) Bad
9.25) All those with excellent LGs
9.26) Use HCl to protonate OH and turn it into an excellent LG
9.29) S_N2
9.30) S_N1
9.31) S_N1
9.32) Neither
9.33) Both
9.34) S_N1

Chapter 10

10.1) Consult your textbook or class notes
10.2) The rate of the reaction is dependent on the concentrations of two compounds
10.3) Consult your textbook or class notes
10.4) The rate of the reaction is dependent on the concentration of only one compound

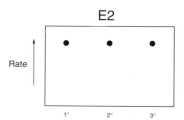

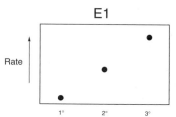

10.5)

10.6) E2
10.7) E2
10.8) E2
10.9) Both

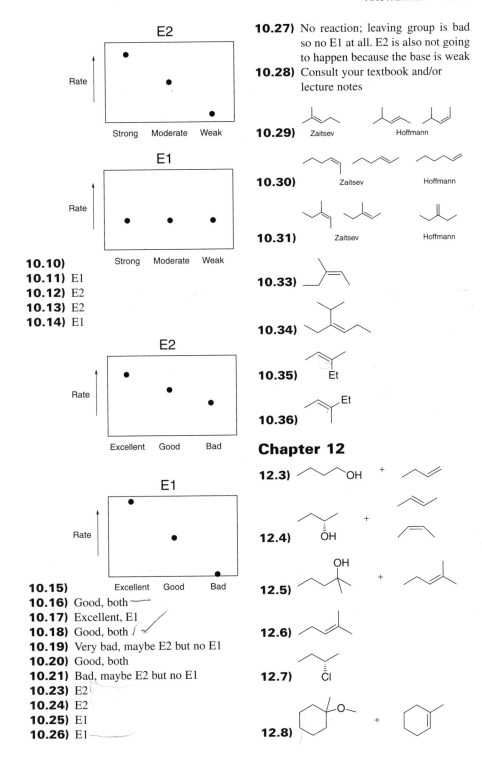

E2

Rate

Strong Moderate Weak

E1

Rate

Strong Moderate Weak

10.10)
10.11) E1
10.12) E2
10.13) E2
10.14) E1

E2

Rate

Excellent Good Bad

E1

Rate

Excellent Good Bad

10.15)
10.16) Good, both
10.17) Excellent, E1
10.18) Good, both
10.19) Very bad, maybe E2 but no E1
10.20) Good, both
10.21) Bad, maybe E2 but no E1
10.23) E2
10.24) E2
10.25) E1
10.26) E1

10.27) No reaction; leaving group is bad so no E1 at all. E2 is also not going to happen because the base is weak
10.28) Consult your textbook and/or lecture notes
10.29) Zaitsev Hoffmann
10.30) Zaitsev Hoffmann
10.31) Zaitsev Hoffmann
10.33)
10.34)
10.35) Et
10.36) Et

Chapter 12

12.3) OH +
12.4) OH +
12.5) OH +
12.6)
12.7) Cl
12.8) O— +

12.9)

12.10)

12.11)

12.12)

INDEX